HOW TO BUILD A HOVERCRAFT

HOW TO BUILD A HOVERCRAFT

Air Cannons, Magnet Motors, and 25 Other **Amazing DIY Science Projects**

CHRONICLE BOOKS
SAN FRANCISCO

Library of Congress Cataloging-in-Publication Data available.

ISBN: 978-1-4521-0952-7

Manufactured in China

Designed by Alissa Faden
Illustrations by Stephen Voltz and Hillary Caudle.
Torus schematic on page 110 by Yassine Mrabet.

DISCLAIMER:
As with any project, it is important that all instructions are followed carefully;
failure to do so could result in injury. Every effort has been made to present the
information in this book in a clear, complete, and accurate manner; however,
not every situation can be anticipated and there is no substitute for your own
common sense. Wear eye protection when advised. Use caution when handling
dangerous objects. Be careful when using power tools or manipulating projectiles.
Before beginning any project, ask yourself what could go wrong, and plan for
that contingency. The authors and Chronicle Books disclaim any and all liability
resulting from injuries or damage caused during the production or use of the
projects discussed in this book.

10 9 8 7 6 5 4 3 2 1

Chronicle Books LLC
680 Second Street
San Francisco, California 94107
www.chroniclebooks.com

CONTENTS

THE COOLEST SCIENCE EXPERIMENTS OF ALL TIME—

THAT YOU CAN DO AT HOME

When a friend of ours first told us about what happens when you drop Mentos into Coke, we had to try it right away. We took a bottle outside, dropped in some Mentos, and were astounded—and inspired.

The next night we put together a ten-bottle geyser performance for *The Early Evening Show* at the late, great Oddfellow Theater in Buckfield, Maine. The audience's response was so enthusiastic that we knew we had to take it further. So over the next several months, we worked nights and weekends developing what turned out to be a 101-bottle Coke-and-Mentos-geyser extravaganza. When the video of *that* went online, it went viral literally overnight. As we write this, we estimate that the video has been seen over 100 million times.

Making that video inspired our next quest: to find the coolest science experiments of all time—that could be done with ordinary everyday objects. We didn't want anything that involved unusual chemicals or special equipment. We wanted amazing things that anyone could do using materials that could be found at home or bought at a neighborhood hardware, office supply, or grocery store.

And we wanted experiments that were FUN. In fact, our ultimate test was whether or not an experiment was entertaining enough to be the centerpiece for a vaudeville-style number at the Oddfellow Theater.

The result is this book.

Then we wanted to add one more thing: *science*.

When we were working on creating our 101-bottle Coke and Mentos spectacular, we spent a lot of time learning the science behind why it works. We tried every soda we could find. We tested every candy on store shelves. We scoured the Internet for information on what made it all explode so gloriously.

Why? We weren't that interested in learning about the science for its own sake, we just wanted to know how to make our geysers go higher! But a strange thing happened as we went through this process. We discovered that understanding why these things work is almost as much fun as doing them. So in this book we try to explain the science behind all the cool stunts as thoroughly as we can.

All our lives we've been the kind of people who collect books like this one. We have a *lot* of them on our own shelves. And almost all of them have either

1) a collection of really great tricks with only a passing nod to the science of how they work, or 2) really thorough scientific explanations for a bunch of rather ho-hum experiments. We wanted a book that included the best of both. So that's what we have tried to write: a book that combines the fun of creating truly cool stunts with the fun of understanding what makes them tick. After all, knowing the science is the best way for you to find ways to take these experiments further, making them bigger, better, even cooler still. It's how you make your geysers go higher.

So roll up your sleeves and get ready to make your own hovercraft. Set off a Coke and Mentos dragster. Make breathtaking steel wool fireworks. Turn soap into a growing mass of lava. And then use the science to figure out your own book's worth of variations. We can't wait to read it.

BUT WAIT—THERE'S MORE!

You can visit our website, EepyBird.com to find additional resources, share your own ideas, watch videos of our experiments, and find even more things that you can try at home. There, you'll find things like our videos of a quarter million sticky notes turned into waterfalls and a Coke-and-Mentos-powered rocket car big enough to propel a human.

Throughout this book, you'll also find Quick Response (QR) codes so that you can use your smartphone to take you directly to more information online. Using a QR reader app, just point your smartphone at the QR code, and your browser will automatically show you the appropriate web page at EepyBird.com.

If you're not sure which QR reader to use, we've got links to a couple of good, safe QR readers at EepyBird.com, and if you don't have a smartphone or QR reader app, don't worry: you can also find all the same fun stuff by visiting EepyBird.com and clicking on "Experiments!"

Now, go get your hands dirty.

AN EXAMPLE OF A QR CODE

STEPHEN VOLTZ
FRITZ GROBE
Buckfield, Maine
September 2013

LEVEL 1

QUICK AND AMAZING

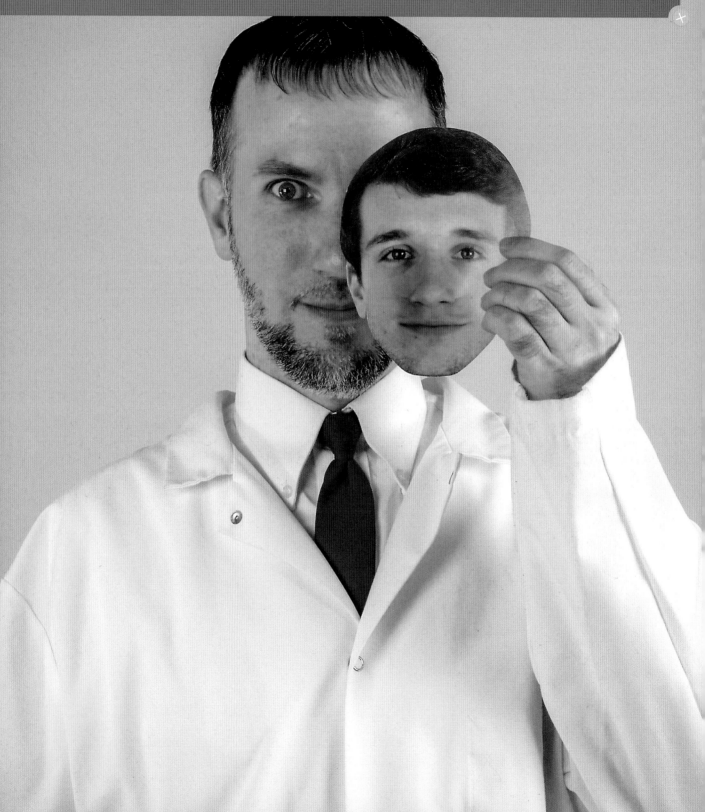

Want to freak out your friends? Create a life-size, nearly realistic head that appears to follow you as you move around the room. All you have to do is take a picture of your face, print it out at life-size, cut it and fold it just the right way, and you can create an optical illusion that appears to turn to watch you no matter where you are in the room. You can also make a surprisingly effective paper "security camera" that seems to follow your every move.

HOW DOES IT WORK?

When you see a face, your brain naturally assumes that the face is convex (sticking out) like a normal face. But the Face that Follows is actually concave (inverted), and that, surprisingly, tricks your brain into thinking the face is looking at you, wherever you stand.

How could your brain get this wrong? In fact, your brain is constantly interpreting the information it gets from your eyes and making guesses about what you're seeing. Is that a real car or a toy car? Your brain examines visual cues and considers the context to decide. How fast is the car moving? Is there a person inside it? Is the car on the road or inside a room? Usually, your brain has no trouble. Even one obvious cue is enough for it to guess correctly.

But convex/concave images can cause us fits. Scientists theorize that the brain makes guesses based on what it's seen in the past. Since most objects in the real world are convex and very few are concave, your brain interprets an ambiguous convex/concave image as convex (unless it has some other information).

This presumption is even stronger with images of human faces, which our brains are inclined to "see" even where they don't exist. All of the many thousands of faces we've seen in our life are convex, so our brains tend to assume that all faces are convex. When you see the hollow mask of the Face that Follows, your brain is convinced that it must be convex. That illusion makes for some unexpected fun.

THE EXPERIMENT: THE FACE THAT FOLLOWS

MATERIALS

- 8½" x 11" white card stock or printer paper
- Scotch tape

TOOLS

- digital camera
- printer (color is preferred, but black-and-white will work)
- scissors

CONCAVE VERSUS CONVEX

If you keep getting confused between *concave* and *convex*, you're not alone. An easy way to tell the difference is to remember that anything *concave* makes a "cave"; it describes an object that curves inward, like a cave or bowl. Its opposite, *convex*, refers to something that curves outward, like a hill or ball. So just remember, the word *concave* contains a clue to its shape, since it has a "cave" inside it.

HOW TO BUILD IT

STEP ONE: Enlist the help of a friend. First, you need to take a digital picture of yourself or someone else. You want a photo with a plain white background if possible, but any uniform light background will do. A white wall is good.

For a clear image without harsh or distracting shadows, use as much light as possible from more than one source. Don't use just your camera's flash. Turn on all the lights in the room, and even take the lamp shades off any lamps, to help make the photo brighter but with a diffused light that creates minimal strong shadows.

Then, while the photographer holds the camera at the subject's eye level, look directly into the lens. This is critical to achieve the illusion. If you look even a little to one side, or a little up or down, your Face that Follows won't be looking directly at the viewer, which is what you want. Take a few photos. Try different expressions. Try a mean face, a mischievous face, a friendly face. We've found that a blank "no expression whatsoever" look makes for a wonderfully unnerving Face that Follows.

The photographer should zoom in so that the subject's face fills the entire frame of the picture. That way when you print the photo, the face fills the whole page. While you can crop and resize the digital image, you avoid this step if the face already fills the entire frame.

NOTE: Smooth down your hair before you take the picture. In STEP THREE, you will cut out the image, and curls or wisps of hair that stand out are hard to cut around.

STEP TWO: Print the picture so that it takes up just about one full sheet of 8½" x 11" card stock or paper. The face of an adult is just about life size when the top of the head to the bottom of the chin is printed on a sheet of 8½" x 11" paper. If your head is a little smaller, fill the whole page anyway. A slightly larger than normal face makes for a great look with this experiment.

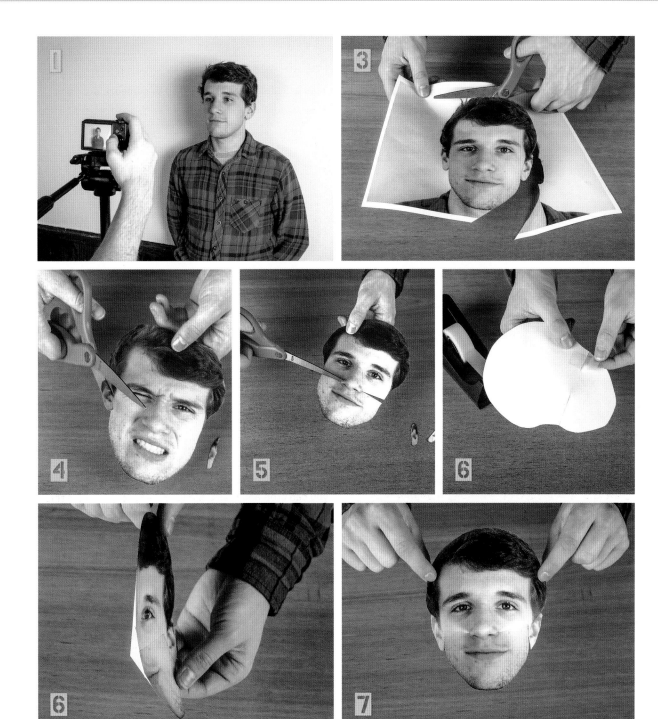

HOW TO BUILD A HOVERCRAFT 13

If you print on white card stock, your Face that Follows will be pretty sturdy. If you use plain white paper, the Face that Follows works just fine, too; it's just more delicate and easier to wrinkle.

STEP THREE: Using scissors, carefully cut out the image of your face, removing all the surrounding white paper. Include your neck and shoulders in what you cut out if they're in the original image. Discard everything but the image of your head, neck, and shoulders.

STEP FOUR: Cut the ears off the image and save them to reattach in STEP SEVEN.

STEP FIVE: Using scissors, carefully make 2 cuts in the photo: one from the base of the left ear to the edge of the left nostril, and then a second cut from the base of the right ear to the edge of the right nostril. Note that no paper is removed in this step.

STEP SIX: On each of the cuts you just made, overlap the edges by about an inch at the ends, so that the image of your face makes a slight bowl. Tape the back of the overlap so the paper holds this bowl shape. You should now have an image of your face that is slightly concave.

STEP SEVEN: Reattach the ears you removed in STEP FOUR so that they are where the ears should be. Your face is done!

STEP EIGHT: Tape your Face that Follows on the wall at least 6' up from the floor. The higher you can place it, the better. The illusion doesn't usually kick in until the viewer is at least 4' or 5' away, so placing it high up on a wall helps ensure that.

NOTE: Before taping, make sure the tape you use won't mark the wall or lift off paint or wallpaper when you remove it.

Now move around the room. The face will seem to turn to follow you wherever you go. Stand on a chair. Lie on the floor. The face will turn up, down, left, and right to look directly at you. If you and a friend both move to different corners of the room, each of you will swear that the face is looking at you even though you're in different places.

THE EXPERIMENT: THE PAPER SECURITY CAMERA

Print out this simple security camera illusion, place it high on a wall in the corner of a room, and watch as people notice that it seems to swing around whenever they walk past it. This may be the most inexpensive security device ever created.

MATERIALS

- the security camera image on this page

  OR

- the security camera image from EepyBird.com (which you can download for free)
- 8½" x 11" white card stock or printer paper (if you're downloading the digital image)
- Scotch tape

TOOLS

- color printer (if you're downloading the digital image)
- scissors

You can find this image at www.eepybird.com/experiments/facethatfollows or use the QR code at left.

HOW TO BUILD IT

STEP ONE: Photocopy the security camera image on page 15 and cut it out. You can also download the security camera image from EepyBird.com and print it out from your computer on white card stock or paper. Carefully cut out the security camera and tabs, and then fold it so that it forms three sides of a box shape.

Remember, to create the illusion, you want to make an *inverted* version of the camera. So make sure to fold the cutout so that the security camera image is on the *inside of the box*, not the outside. This is the opposite of what you'd do ordinarily. Tape the folds on the outside, unprinted side of the paper, so that the box holds its shape.

STEP TWO: Find a good spot to place your "camera." You want to put your fake camera high on a wall where it will appear to look down on people. An ideal location is up against the top corner of the room, placing it flush where the two walls meet the ceiling (see illustration).

Use a few small loops of tape on the back to hold the camera in place. Before doing this, make sure the tape you use won't mark the wall or lift off paint or wallpaper when you remove it.

Walk around the room and watch the camera appear to move. It will seem to move both side to side and up and down, matching your movements exactly, as if someone is operating the camera remotely and moving the camera to look around the room. It's wonderfully creepy.

Print up 5 or 6 and put them up wherever you think a fake robotic camera might be a good prank. We're not telling you to put one out in front of your local police station, but if you do, send us a picture!

NOTE: The farther away the fake camera is from people, the better the illusion. It begins to work pretty well at 6' or 7' away. Closer than that and it's not as effective. If you're having trouble getting it to work for you, close one eye and move around the room, looking at the camera that way. Once you get it working with one eye closed, you can often open both eyes and it will continue

to work. Also, the illusion works extremely well on video. A cell phone recording of the camera appearing to follow you around can look surprisingly real.

The illusion created when we see a concave shape and think we're seeing a convex shape has been known for over two hundred years. It occurs because, while we live in a three-dimensional world, each of our eyes can see only in two dimensions.

A single human eye works in the same basic way that a camera or a movie projector does. A lens focuses the light that enters on one side and projects a flat two-dimensional image onto the other side. While a projector focuses its image on a screen and a camera focuses its image on an electronic image sensor (or in the old days, film), the lens of our eye focuses its image on the retina on the back of the inside of our eye. Each eye then sends that image to our brain through the optic nerve, and our brain interprets the information and tells us what we're seeing. For example, the slight differences between the images that our right and left eyes send to our brain help us determine things like depth. When you see a 3-D movie, the special glasses you wear allow each eye to see a slightly different image, creating the sense of depth.

Most of the time, our brain interprets all this information with no trouble, but two-dimensional images can be ambiguous. Once in a while our brain has to guess what we're actually seeing, choosing among the different possibilities of what our two eyes are showing.

Perhaps the simplest example of this was discovered by Swiss scientist Louis Albert Necker, who in 1822 published an illustration of what is now widely known as the "Necker Cube": this drawing can be viewed either as a cube coming forward and facing down and to the left, or as a cube coming forward and projecting up and to the right.

The problem with the two-dimensional information we get from our brain isn't limited to the Necker Cube. For example, the image our eye gets from a line pointing toward us is nearly identical to the image our eye gets from a line pointing away from us. In the same way, a convex face creates the same two-dimensional image on our retina as a concave face.

Thus, from straight on, a convex shape looks exactly the same as a concave shape, but seen at an angle, it adopts the opposite perspective: from the right, the convex shape creates the same two-dimensional image as the concave shape would from the left (see illustration).

As a result, as the viewer moves from left to right, a concave (inside-out) face looks the same as a convex (normal) face would if that face were turned directly to the right and toward the viewer. This works in the reverse direction: as the viewer moves from right to left, a concave face seems to turn directly to the left and toward the viewer.

With the security camera, from whatever angle we view it, the two-dimensional image that the inside-out camera creates on our retina is exactly the same as the image a normal camera would create if it were pointing right at us. Once our brain has decided that what we're seeing is a convex shape, our brain continues to interpret that object as convex as we move around.

Different Objects Can Create the Same Image on the Retina

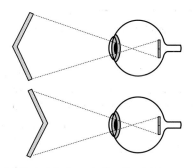

An object pointing forward and an object pointing away both project the same image on the retina

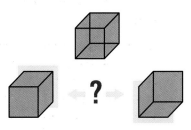

The eye can't tell if the cube on the top is facing up like the cube on the right or down like the cube on the left

THE ILLUSION DOESN'T WORK ON BRAINS THAT AREN'T WORKING RIGHT

Interestingly, psychologists have discovered that people whose brains are affected by schizophrenia aren't fooled by these kinds of illusions, nor are people who are drunk or under the influence of certain drugs. No one is sure yet why this is, but it appears to be directly related to the fact that people with these conditions sometimes hallucinate and convincingly "see" and "hear" things that aren't there.

In both situations—where people are hallucinating and seeing things that no one else does, and where people are failing to see optical illusions that others do—there seems to be a disconnect in the brain between the signals that the eyes and ears are sending and how the brain is interpreting those signals. The fact that people suffering from schizophrenia can't see certain optical illusions may turn out to be a key to understanding what's going on in their brain circuitry and causing their disease.

For more ideas, videos, and variations, visit www.eepybird.com/experiments/facethatfollows.

This same basic phenomenon occurs with most optical illusions. Once they "click in," we see them easily, and it can be difficult to see them any other way. Then once they "click out," we can't see the illusion at all, even though we know it's there. This is called "*gestalt* switching," from the German word *gestalt*, which means (roughly) "the entire thing."

Both the Face that Follows and the Paper Security Camera trick the brain into guessing (incorrectly) that what we're seeing is convex. Then, as we move around, our brain guesses (incorrectly) that the convex image we're seeing is also moving and turning with us, since that's what would be happening if the object was actually convex.

The Growing Head Illusion is one of the most mind-bending optical illusions ever discovered. Simply gaze at the center of the spinning spiral for 20 seconds, then look at your friend, and it will seem as though his or her head is expanding like a giant balloon. The illusion typically lasts for as long as 10 to 15 seconds.

Reverse the direction of the spiral, and your friend's head will seem like it's shrinking.

HOW DOES IT WORK?

Sometimes known as the "waterfall illusion" or the "motion aftereffect," this illusion is said to have been observed by Aristotle over two thousand years ago. An Englishman, R. Adams, named it the "waterfall illusion" after observing the effect while watching the Scottish "Fall of Foyers." Whenever he looked away from the waterfall after observing it for some time, the rocks to the side of the falls appeared to be moving upward.

In this experiment, as you gaze at the spinning spiral for about 20 seconds, you experience the illusion that things are getting smaller or fading off into the distance. When you look away, the motion aftereffect reverses the impression, making whatever you look at appear to be growing.

SENSORY ADAPTATION

Exactly what's going on between the eye and the brain to make this happen isn't completely understood. One theory is that the neurons in our eyes become fatigued watching the same motion for too long, and so they are too tired to adjust back immediately when we look away. Another, probably better, theory is that this is a case of what's known as "sensory adaptation."

Almost all of our senses respond to a continuous stimulus by gradually reducing whatever sensation we are experiencing. This is why bathwater that feels too hot at first becomes the perfect temperature if we just stay in for a minute or so. The water remains the same temperature, but our senses adapt. Then, if we're splashed with room-temperature water, that water will feel cold.

In a similar way, a quiet room can suddenly seem impossibly quieter when the refrigerator cycles off. And it accounts for why, upon returning to land after being at sea, ship passengers often feel as though the ground is rolling like waves.

In each of these situations, we first notice a stimulus distinctly, even intensely, but if it remains constant, then it gradually fades out of our consciousness. That stimulus becomes normalized, and we notice any change only in relation to it. If our senses have adapted to a hot bath, room-temperature water will feel cold, and so on.

As you stare at the spinning spiral, your vision grows accustomed to the sensation that everything is shrinking into the distance at the center of the spiral. Then, when you switch to looking at something normal, it looks like it's getting bigger simply because you have adapted to expect that what you are seeing is shrinking.

THE EXPERIMENT: THE GROWING HEAD ILLUSION

MATERIALS

- the spiral image in this book (pages 22 and 23)

  OR

- the spiral image from EepyBird.com (which you can download for free)
- 2 sheets of 8.5" x 11" white paper (if you're downloading the digital image)
- Scotch tape or glue
- cardboard
- duct tape
- electric drill

TOOLS

- color or black and white printer (if you're downloading the digital image)
- scissors

HOW TO BUILD IT

STEP ONE: Photocopy the 2-page spiral design shown on pages 22 and 23. You can also download the 2-page pattern from EepyBird.com and print it out. The spiral will be 9½" in diameter, which is too big to fit on one standard sheet of paper, so each page only shows part of the spiral. Cut out the portion of the whole image on each page, and use Scotch tape or glue to fasten them together to form a circle.

STEP TWO: Trace the circle's diameter on a piece of cardboard and cut it out.

STEP THREE: Glue or tape the paper spiral to the cardboard circle to give it a solid backing.

STEP FOUR: Use duct tape to attach the back of the cardboard spiral to the end of the drill; tape right onto the end of the drill, without using a drill bit. Make sure that the center of the spiral is centered on the tip of the drill.

STEP FIVE: Hold the trigger of the drill so that the spiral spins at a medium speed.

STEP SIX: Have a friend (or better, a group of friends) stand 10 or 15 feet away from the spiral and gaze at the center of the circle steadily for 20 seconds while the spiral spins. Be patient and have everyone gaze steadily the entire time; don't rush. It helps if the person holding the drill slowly counts down from 20.

STEP SEVEN: After 20 seconds, have your friend(s) look directly at the nose of the person holding the drill, whose head will look like it's inflating like a giant balloon. The illusion will last for a full 10 to 15 seconds.

If you repeat this with the drill spinning the other way, it will create the opposite effect: it will look like the person's head is shrinking.

that we might otherwise miss. For example, when we're driving down a highway, we adapt to our speed, and this improves our ability to notice small speed increases or decreases, which without that adaptation we would not be able to notice. From the side of the road, it's hard to tell whether or not a passing car is accelerating from 60 miles per hour to 70 miles per hour, but if you are in that car or driving along next to it, this is easy to tell.

THE POWER OF SENSORY ADAPTATION

Sensory adaptation can work in remarkable and complex ways. For instance, experimental psychology pioneer George Stratton invented a headset with a pair of glasses that made the world appear both upside down and reversed left and right. He wore the glasses, which were rigged with mirrors that inverted the light rays hitting his retina, for about a week, at which point Stratton reported that his brain had adjusted to them, automatically adapting the odd input so that the world appeared perfectly right side up to him. He got around just fine. This was true despite the fact that he knew the glasses inverted the world.

After he removed the glasses, it took several hours for his sensory perception to return to normal. The world no longer appeared upside down and left to right, but he continued to react as if it was. He found himself reaching for objects on his left side with his right hand, and vice versa.

For more ideas, videos, and variations, visit www.eepybird.com/experiments/growinghead.

MULTIPLE ILLUSIONS

The Growing Head Illusion is particularly powerful because it involves multiple optical illusions at the same time. First, the spinning spiral isn't really growing or shrinking. That effect itself is an optical illusion, the simplest version of which is known as the "barber pole illusion."

The spiraling stripes on a spinning barber pole give the illusion of up or down motion (depending on which way the pole spins). As with most optical illusions, the barber pole illusion is based on ambiguous information that our eye receives. The moving stripes on a barber pole create the same image on our retina whether the pole is 1) spinning or 2) not spinning but moving upward. Given this ambiguous information, our brain makes a guess and interprets the action as upward motion.

A spiral creates a circular version of the barber pole illusion. The spinning disc appears to be moving in on itself when spinning in one direction and moving outward and expanding when spinning in the other direction. This gives the illusion that the spiral is growing or that it's shrinking. Sometimes this can also feel as though we're moving in toward the center of the spiral or being pulled back away from it.

When we look away, sensory adaptation creates a reverse illusion: if the spiral appeared to be shrinking, whatever we look at next will appear to grow larger. Thus, a friend's head will seem to inflate like a balloon.

ANOTHER EXAMPLE OF SENSORY ADAPTATION

There's a great demonstration of sensory adaptation using touch at San Francisco's Exploratorium museum involving three copper pipes: one warm, one room temperature, and one cold. In the experiment, you wrap one hand around the warm pipe and the other around the cold pipe.

After about 15 seconds, long enough to establish a new (but different) "normal" for each hand, you wrap both hands around the room-temperature pipe. The room-temperature pipe will now feel astonishingly cold to the hand that had been holding the warm

pipe, and it will feel warm to the hand that had been holding the cold pipe.

This experience has nothing to do with the "objective" temperature of the pipes, or with your conscious expectation of what should feel warmer or colder. Your brain automatically and involuntarily interprets the *change* or difference in temperature and decides what's "warm" and what's "cold." The fact that the room-temperature pipe is experienced as both "warm" and "cold" at the same time, but by different hands, makes clear that the relative difference of change is what guides our sensory perceptions.

To put it differently, our sensing organs—our skin, ears, eyes, nose, tongue, and so on—continue to report accurately to our brain, but the receptor cells in our brain report and notice the *relative* change.

Sensory adaptation like this improves our ability to discern slight but critical changes in our environment

Balloons and needles don't mix, right? What if you could pass a long sharp needle completely through a balloon without popping it? It looks impossible, but it's easy once you know the trick.

HOW DOES IT WORK?

The rubber of a latex balloon is made up of long chain-shaped molecules that curve and fold back on themselves. Materials like this are called *elastomers*. When these elastomer chains are pulled, the curves in them straighten out, allowing them to stretch considerably before the chains break. This is what makes rubber stretchy.

Before a balloon is inflated, the rubber's chain-shaped molecules are in their relaxed state, with numerous curves, twists, and folds. As the balloon is inflated those molecules are pulled apart and stretched out.

When the balloon is inflated, most of the molecules get stretched to their limit, except at two places. That's the key to this trick. At the end and at the mouth, the latex is thicker, and the long, kinked chains remain relatively relaxed. Here, the material can still bend and stretch out of the way of the needle when the sharp point pushes through it.

On the other hand, when the needle pokes the tightly stretched latex on the side of the balloon, there's no more give in those molecules, so the balloon will pop.

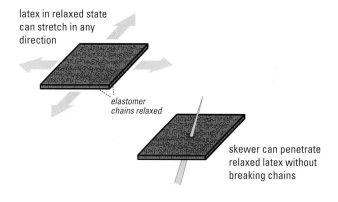

latex in relaxed state can stretch in any direction

elastomer chains relaxed

skewer can penetrate relaxed latex without breaking chains

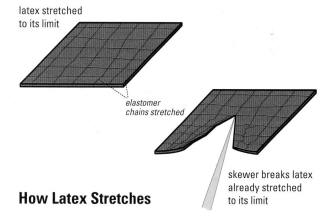

latex stretched to its limit

elastomer chains stretched

skewer breaks latex already stretched to its limit

How Latex Stretches

THE EXPERIMENT: THE NEEDLE THROUGH THE BALLOON

MATERIALS

- 1 "O" gauge (2 mm) or thinner knitting needle or 1 barbecue skewer (steel or bamboo)
- multiple latex balloons ("jewel tone" balloons are best; they are partially transparent and allow you to see the needle inside)
- a few drops oil (vegetable oil or 3-in-1 lubricating oil) or Vaseline

TOOLS

- sandpaper or grinding stone (to sharpen the needle, if necessary)
- wire cutters (if the needle has a head)
- rag or paper towel (to lubricate the needle)

HOW TO BUILD IT

STEP ONE: Make sure the needle is as sharp as possible. If necessary, use sandpaper or a sharpening stone to get the point as sharp as a sharpened pencil. Using wire cutters, snip off the head of the needle if it has one.

After sharpening, make sure that there are no burrs or splinters anywhere on the needle. You want a sharp point and a smooth surface. Be very careful with the needle now! It's sharp!

STEP TWO: Inflate the balloon to about 80 percent of maximum. It should be nicely inflated, but with room to inflate it a little more. With experience you'll be able to inflate your balloons almost completely.

NOTE: You only need 1 balloon to do this, but you'll definitely want to do it a few times. Since you can only do this once per balloon, and you may not get it right the very first time, make sure to prepare multiple balloons through STEP THREE.

STEP THREE: Loosen your grip on the mouth of the balloon very briefly to allow a small puff of air to escape so that the balloon deflates slightly. This helps ensure that the balloon will have enough stretch left in it for the needle to go through without the balloon bursting. Tie off the balloon.

STEP FOUR: Apply a few drops of oil or a spot of Vaseline to the rag and slide it over the needle to lubricate the entire length.

STEP FIVE: Notice that at both the mouth of the balloon where you blow into it and at the opposite side (the very tip of the balloon) there are two small areas where the rubber is thicker and darker than everywhere else on the balloon. These two spots are where you will pierce the balloon.

STEP SIX: Holding the balloon in one hand and the needle in the other, carefully and slowly push the sharp end of the needle into the thickest area of rubber at the end of the balloon near the tied-off mouth. If the balloon is not too fully inflated and the needle is sharp, the needle will pierce the balloon rubber without popping the balloon.

Continue pushing the needle through the balloon, aiming the point so that it hits the thickest part of the rubber on the opposite side on the way out. If you hit the right spot, the needle will pierce the rubber again and exit the balloon without breaking it.

STEP SEVEN: After letting your audience enjoy this spectacle for 5 to 10 seconds, you can gently remove the needle from the balloon by continuing to pull it out in the same direction that you pushed it in. As soon as you remove the needle, air will begin to leak out, slowly but noticeably, from the 2 holes the needle made. At this point, after appreciating the intact balloon, we like to demonstrate that both the balloon and the needle are real by poking the sharp end of the needle into the side of the balloon, where the rubber is thinnest, and popping the balloon. Don't wait too long to do this, however, or the balloon may not be inflated enough to pop easily.

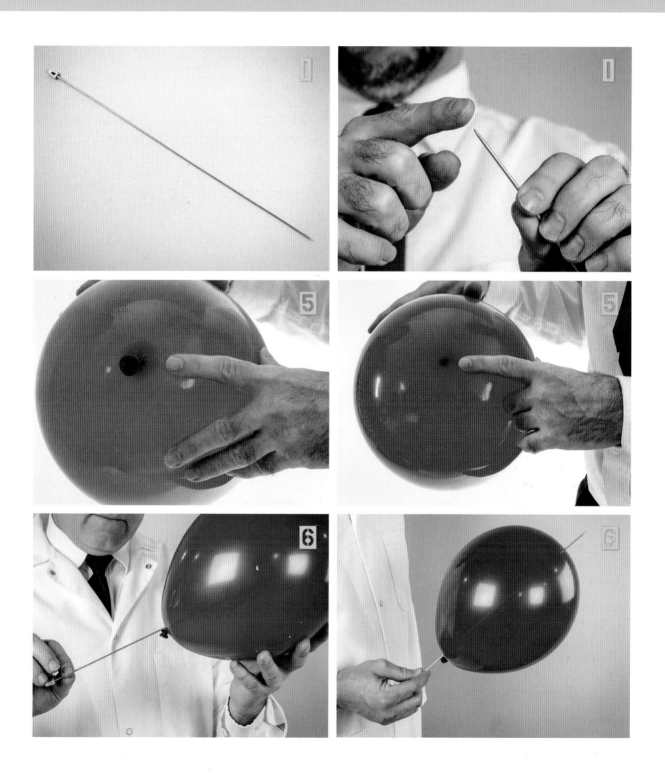

THE SCIENCE

As we said earlier, this trick depends on sliding the slippery needle through the balloon in those two spots where the latex isn't stretched as much as it is elsewhere on the balloon. Why do balloons have these two thicker spots?

USING THE BALLOON'S SHAPE

Latex balloons are manufactured by dipping a metal form that is the size and shape of an uninflated balloon into liquid latex, which covers the form and coats it. Brushes then roll back the rubber at what will become the mouth of the balloon to form the classic balloon "lip." The rubber is then heated to cure it, and the balloon is removed from the form.

However, while the latex is still liquid, some of it drains down from the long neck section and collects around the top, and it also drains from the sides and collects at the bottom in the area that will become the top of the balloon. This creates two areas on the balloon where the latex is thicker than everywhere else. These are the two spots where the needle can pierce the balloon without breaking it.

WHAT'S THAT POWDER?

Why, you wonder, doesn't the balloon stick to the form when it's made? That's because prior to dipping it in the liquid latex, the form is covered with powder in the same way that you might powder a cake pan before pouring in cake batter. Have you ever noticed this powder when you've blown up balloons in the past? That's what this is.

THE MAGIC OF RUBBER

Rubber is an amazing material. It's made from latex, the milky saplike liquid (though it is not itself sap) that exists naturally in certain plants. Most commercial latex these days comes from the Pará rubber tree, though latex itself is quite common and can be found in some amount in about one in ten flowering plants in the world. And while "latex" is technically the raw

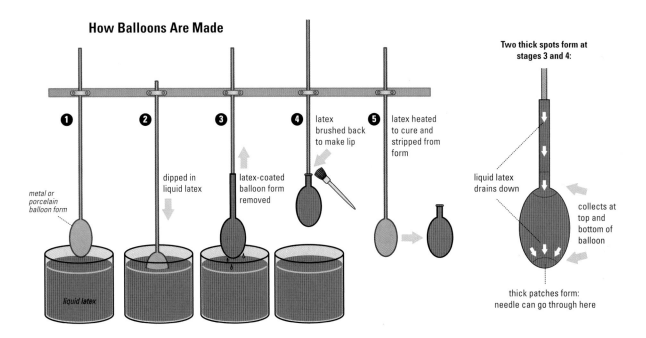

How Balloons Are Made

① metal or porcelain balloon form

liquid latex

② dipped in liquid latex

③ latex-coated balloon form removed

④ latex brushed back to make lip

⑤ latex heated to cure and stripped from form

Two thick spots form at stages 3 and 4:

liquid latex drains down

collects at top and bottom of balloon

thick patches form: needle can go through here

plant material and "latex rubber" is the end product, the word "latex" can mean either one.

Latex is tapped from rubber trees in much the same way maple trees are tapped for the sap that is boiled down into maple syrup.

After harvesting, the liquid latex is cured by heating it, which solidifies it and makes it less sticky. This is the material we use to make things like balloons and rubber gloves. When rubber is "vulcanized" by heating it with sulfur and other stabilizers, it becomes even sturdier and more stable, and it can be molded into things like rubber balls, waterproof boots, and automobile tires.

Rubber and the process of vulcanization was discovered over thirty-six hundred years ago by the Olmec people—predecessors to the Aztecs and Mayans—of Central America. The Olmecs (whose name itself means "rubber people" in Nahuatl, the language of the Aztecs) tapped a Central American variety of rubber tree for latex, mixed it with juice from the sulfur-rich moonflower morning glory, boiled it down, and created a sturdy, moldable rubber that they used to make solid rubber balls, hollow human-shaped figurines, and strong, thick rubber bands that they used, among other things, to hold axe heads onto axe handles. Central American peoples also used this rubber to make sturdy rubber soles for their sandals and rubber tips for specialized hammers and for the heads of certain drumming mallets.

When rubber was first introduced to the West in the 1700s, Europeans did not appreciate its practical potential. They did find it useful as an eraser for rubbing off the marks made by a pencil, however, which is how the substance came to be known as "rubber."

For more ideas, videos, and variations, visit www.eepybird.com/experiments/needleballoon.

In this experiment, you take a 1-gallon metal can and cause it to implode dramatically. All you need to do is heat up some water inside the can, seal it up, and then cool everything down. That is enough to make the can crumple and collapse, seemingly all by itself.

HOW DOES IT WORK?

The can is actually crushed by the invisible air pressure that's all around us on Earth. So, what exactly is air pressure?

Air may seem like nothing at all, but it's made up of gases, nitrogen (about 80 percent), oxygen (about 20 percent), and a little bit of argon (less than 1 percent). And while, relatively speaking, all gases are very light, there's a lot of air above and all around us. At sea level, there's about one ton of pressure pushing down on us all the time—just from the weight of the atmosphere.

But air pressure doesn't just push down. The air molecules around us are always moving and pushing in all directions, up, down, and sideways. That means that so long as the can is open, and the air can flow in and out, the pressure inside the can (molecules pushing out against the wall of the can) will balance out the pressure outside the can (the molecules pushing in from the outside). This equalizes the pressure, which then causes no noticeable effect at all.

In the Self-Crushing Can experiment, you crush the can by changing the pressure inside the can so that it becomes much weaker than the air pressure surrounding it. How? First, you boil water inside the can, filling it with steam. The heat gets all the molecules inside the can moving around and the gases expand, taking up more space. Then you seal the lid, so that nothing can flow in or out of the can, and you quickly cool down the steam, which condenses back into liquid water. This creates a vacuum inside, since water molecules take up far less space as a cool liquid than they do as a steam. With the water molecules inside the can condensed back to liquid, the air molecules outside the can have nothing pushing back against them, and they crush the can.

Notice here that we're used to thinking of a vacuum as a force, like the sucking force of a vacuum cleaner. But in fact, a vacuum isn't a force in itself. It's really the *absence* of pressure. What crushes the "self"-crushing can isn't the vacuum you create inside by cooling the steam, it's the force of the air molecules *outside* the can, which continue pushing against it like always—but suddenly have nothing pushing back against them.

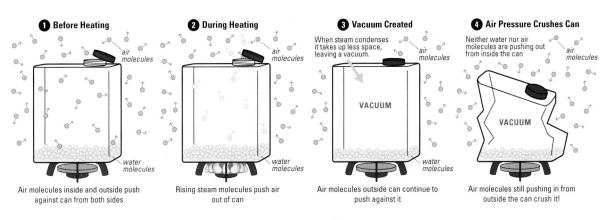

1 Before Heating
air molecules
water molecules
Air molecules inside and outside push against can from both sides

2 During Heating
air molecules
water molecules
Rising steam molecules push air out of can

3 Vacuum Created
When steam condenses it takes up less space, leaving a vacuum.
air molecules
VACUUM
water molecules
Air molecules outside can continue to push against it

4 Air Pressure Crushes Can
Neither water nor air molecules are pushing out from inside the can
air molecules
VACUUM
Air molecules still pushing in from outside the can crush it!

THE EXPERIMENT: THE SELF-CRUSHING CAN

MATERIALS

- one empty 1-gallon "F-Type" metal can (These are the kinds of cans that solvents like paint thinner and mineral spirits often come in.)
- water
- ice

TOOLS

- safety goggles
- measuring cup
- stove or hot plate
- 2 oven mitts
- basin of ice water large enough to hold the can

HOW TO BUILD IT

STEP ONE: Make absolutely certain that the can is completely empty of all combustible liquids. Pretty much any liquid that comes in a can like this is combustible. If there is a small amount of leftover solvent in the can, transfer it to a safe, flameproof container with a tight lid, such as a Mason jar. (Don't use plastic containers since the solvent might dissolve and melt the plastic.) Wash the can out with water and let it air dry with the lid off for a couple hours to let any remaining solvent evaporate. You're going to be heating this can up on the stove and the last thing you want to be doing is heating anything combustible.

STEP TWO: Put on safety goggles. The can may crush very dramatically, even instantly. This explosiveness can have unexpected results, so protect your eyes from any accidentally flying objects. Set the basin of ice water near the stove. Alternatively, if your sink is near the stove, you can fill it with ice water instead of using a basin.

STEP THREE: Pour approximately ¼ cup of water into the can. You want to cover the bottom of the can completely, but the can should still be nearly empty.

STEP FOUR: With the lid or cap *off*, put the can on the burner and heat the water until it boils. While you wait for the water to boil, put oven mitts on both hands.

STEP FIVE: Once the water is boiling fully, and creating lots of steam, turn off the burner.

STEP SIX: Using the oven mitts, lift the can off the burner, turn it upside down, and immediately place it, still upside down, into the ice water. The can will almost immediately crush itself quite forcefully and dramatically.

NOTE: Turning the can over and quickly putting it in the ice water upside down ensures that the water seals the opening in the can so that no air can enter to equalize the pressure.

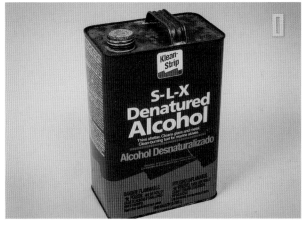

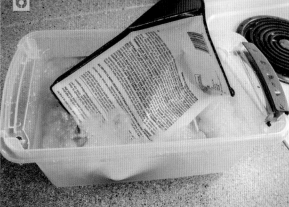

THE SELF-CRUSHING SODA CAN: A QUICK AND EASY VARIATION

You can also do this same experiment with an ordinary soda can. Watching a large steel can instantly crushed by the atmosphere is pretty impressive, but if you don't have a steel can, try it this way.

MATERIALS + TOOLS

You'll need all of the same tools and materials as you did for the initial experiment except for the measuring cup. You'll also need a pair of kitchen tongs, and instead of the F-Type metal can, use a 12 ounce aluminum soda can.

HOW TO BUILD IT

STEP ONE: Fill a soda can with a couple tablespoons of water.

STEP TWO: Place the soda can with the water in it on the burner and heat until the water is at a full boil. Put on your oven mitts.

STEP THREE: When the water is boiling, using the tongs (the can will be too hot to touch), lift the can off the burner and immediately turn it upside down and put it, mouth down into the basin of ice water. CRUSH!

THE SCIENCE

If all this air pressure is strong enough to crush a metal can, why don't we notice it all of the time? Why doesn't it crush *us*?

It's only when we create an imbalance between the pressure inside and outside the can that the air pressure crushes it. Our bodies have air inside (in our lungs, for example) that is under equal pressure as the air around us. So long as our bodies keep that pressure balanced, we don't have any problems. We're used to a certain amount of air pressure.

However, when we fly in an airplane, high up in the atmosphere, there's less pressure. We may find our ears popping as our body tries to balance the pressure inside and outside of itself. And if we drink from a plastic bottle and seal it back up while the plane is high in the air, we will find, upon landing, that the bottle is crushed just like the can in this experiment. All because the air pressure is higher on the ground than it is way up high.

If we subject our bodies to a dramatic change in pressure too quickly, we can experience what's called decompression sickness, known to divers as "the bends." When we dive deep underwater, we have all that atmosphere pushing on us, plus, we have a lot of much heavier water pushing on us as well. When deep divers surface too quickly, the changing pressure can cause gases in their blood to form bubbles with painful and dangerous consequences.

Astronauts also need to be concerned about air pressure and possible decompression sickness when they move between inside and outside a space capsule. No one wants to be inside a crushing or bursting can when the relative air pressure changes.

And no one wants to be the can.

For more ideas, videos, and variations, visit www.eepybird.com/experiments/selfcrushingcan.

A simple, ordinary bar of Ivory soap can grow lava-like into a giant mound of pure white foam in a matter of seconds, right before your eyes. All it takes is some Ivory and a microwave oven.

HOW DOES IT WORK?

Why does a bar of Ivory work for this experiment, while many other brands of soap do not? Ivory soap was first put on the market by Procter & Gamble in 1879, and since 1891 it has been marketed with the catch phrase "It Floats!"

Originally, floating soap had an advantage over other soaps because the bar would not disappear to the bottom of the sink or bath (or as in the ad below, the creek). Instead it would always remain on top of the water, where it could be easily found if dropped.

Ivory floats because it's whipped during manufacturing, creating tiny pockets of air that make the soap less dense than water. It's all those tiny pockets of air that turn it into soap lava in the microwave. As you'll see, the moisture in those pockets causes them to expand like balloons when they're heated, and that makes the entire bar turn into a mass of frothing, expanding foam.

An Ivory Soap ad from 1898: "It Floats!" Ivory Soap, Strobridge & Co. Lithograph, 1898, restoration © Adam Cuerden, used by permission.

INVENTING FLOATING SOAP

Legend has it that the folks at Procter & Gamble discovered floating soap when a workman accidentally left an Ivory mixing machine on too long during the manufacturing process. The tiny air bubbles that were whipped into the liquid made a batch of soap so light that the resulting bars floated in water. Customers who used these bars raved about them, so the company began making Ivory that way on purpose.

Though it's a fun story, it's completely made up. In fact, Procter & Gamble chemist James N. Gamble, son of company cofounder James Gamble, had previously learned how to make floating soap in his study of soap making, and he consciously decided that all Procter & Gamble soap would be made that way well before the Ivory soap brand even existed. How do we know this? In his 1863 notebook, the younger Gamble wrote, "I made floating soap today. I think we'll make all our stock that way."

THE EXPERIMENT: THE IVORY VOLCANO

MATERIALS

- 1 bar of Ivory soap

TOOLS

- sharp knife
- microwave-safe plate (a paper plate works, too)
- microwave oven

HOW TO BUILD IT

STEP ONE: To start, use a sharp knife to carefully cut your bar of Ivory in half. You can use a full-size bar, but to start it's better to use a half bar. You'll be surprised how big just a half bar gets.

STEP TWO: Place the Ivory on the plate and in the microwave.

STEP THREE: Set the microwave to "high" and microwave the bar for 30 to 45 seconds and watch as the Ivory heats up. At first, the bar will just sit there, but after about 10 seconds, you should notice a large white lava-like eruption oozing out at a good clip. As the bar continues to heat, the plate will quickly fill with the white soap lava. Depending on the strength of your microwave, it may need a little less or a little more time to fully expand. Stand by and be ready to stop the microwave, because even 30 seconds may be enough time for the soap to start overflowing the plate.

STEP FOUR: After 30 to 45 seconds, turn off the microwave and open the door, but do not remove the soap lava. Instead, let it cool first. Notice that it will deflate a bit and shrink slightly in size as it solidifies.

STEP FIVE: Once the lava has cooled, after about 15 seconds, remove it from the microwave. It will have transformed back into a solid. You can break it up, put it in a small bowl, and use it as decorative soap. You can also crush it easily. It makes a very realistic fake snow for Christmas displays.

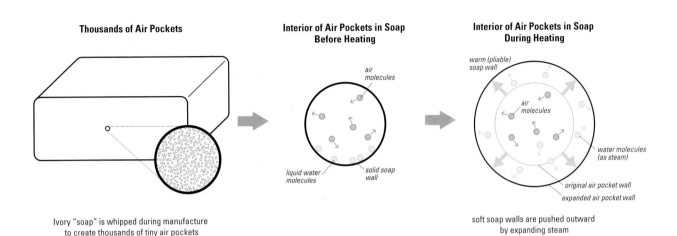

Thousands of Air Pockets

Ivory "soap" is whipped during manufacture to create thousands of tiny air pockets

Interior of Air Pockets in Soap Before Heating

air molecules

liquid water molecules

solid soap wall

Interior of Air Pockets in Soap During Heating

warm (pliable) soap wall

air molecules

water molecules (as steam)

original air pocket wall

expanded air pocket wall

soft soap walls are pushed outward by expanding steam

The fact that Ivory has tiny air bubbles whipped into it is the reason it turns into soap lava in the microwave. The pockets of air in Ivory make the soap a type of *foam*. Foam is formed by trapping pockets of gas into a liquid or solid. Foams are all around us. A kitchen sponge is a foam, the whitecaps on ocean waves are a foam, a cheese soufflé is a foam, and even Swiss cheese itself—full of all those distinctive holes—is a foam.

In Ivory soap, each tiny hole is filled with air that has a little bit of moisture in it. When you microwave a bar of Ivory, that moisture turns to steam and expands. At the same time, the Ivory itself softens and becomes pliable because the heat adds energy to the soap molecules, which move around more. The solid bar starts to liquefy as the soap molecules become less firmly attached to their neighboring molecules.

The heated soap walls become soft and pliable and the steam-filled pockets of air inflate like thousands of tiny balloons. As these tiny expanding soap balloons grow, your bar of Ivory bubbles like an erupting volcano.

Next, when you turn the microwave off and stop heating the bar, two things happen:

First, the soft foam stops growing and even deflates a little. Why? Is air leaking from the foam?

Nope.

What happens is almost exactly the same as what happens on a larger and more dramatic scale with the Self-Crushing Can (page 34). The steam in each tiny air pocket cools and turns back into water, creating a vacuum. Liquid water takes up less space than steam (a gas), and so each air pocket now has a lower air pressure than the air outside the soap. As with the Self-Crushing Can, the air molecules outside push in against a vacuum, causing a collapse.

Second, as the soap cools, the molecules slow down (which is actually what "cooling" means). The soap hardens, but now in the shape of the lava. The soap itself is actually just as hard as it was in bar form, but its structure is much weaker, since it is now filled with much larger air pockets separated by thinner soap walls. The soap molecules haven't changed, but the structure of the bar has. Thus, the lava you created is much easier to crumble and break apart, but you can still wash with it.

For more ideas, videos, and variations, visit www.eepybird.com/experiments/ivoryvolcano.

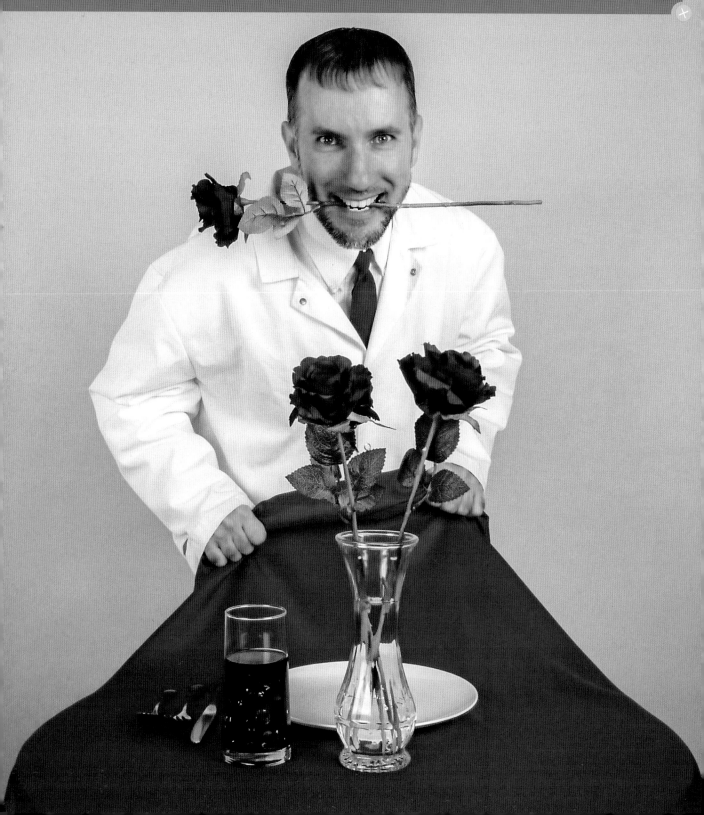

Here, you start with the classic whip-the-tablecloth-out-from-under-the-place-setting stunt. It never gets old, and we'll show you how to do it. Then, you scale it up into even more impressive tricks—using eggs, glasses full of water, and finally, a chain reaction of dropping golf balls.

HOW DOES IT WORK?

All of these stunts work on the same principle. A large-enough force applied to the bottom object of a stack can displace that object without disturbing the other objects on top of it. It's inertia versus friction: that is, the tendency for objects at rest to stay at rest versus the frictional force of two objects sliding against each other. It's odd that removing something quickly, like a tablecloth, can have such a different effect than removing it slowly, but as you'll see, speed here really makes all the difference.

WARNING!

The tablecloth trick takes practice to do without breaking anything! There is a learning curve! For this reason, don't use any items that you (or your family) would mind breaking.

Further, the materials make a difference. Real tablecloths usually have a hem, which will catch and dislodge the objects on the table when the cloth is pulled under them. You must use unhemmed, smooth fabric for your "tablecloth." The easiest solution is to buy a smooth fabric remnant at a fabric store or at the fabric section of any large department store. You could also cut off the hem along one edge of a tablecloth your family no longer uses (and make sure this unhemmed edge is placed at the back for your trick).

Also, all the dinnerware and tabletop items should be as heavy as you can find. Consider practicing first with unbreakable items, then adding ceramic and glassware once you've got the hang of it.

THE EXPERIMENT: THE CLASSIC TABLECLOTH TRICK

MATERIALS

- An unhemmed cloth large enough to cover your table (NOT an actual tablecloth)
- A small table, no larger than 30" deep
- A variety of heavy, smooth-bottomed plates, saucers, water glasses, wineglasses, silverware, wide-bottomed vases, and fruit bowls (NOT your family's regular dinnerware)
- Silicone spray lubricant (optional; available at hardware and auto-supply stores)

HOW TO BUILD IT

STEP ONE: Lay your tablecloth over the tabletop so that it's closer to you than normal: only about 2 feet of cloth are on the table and the rest hangs over the side closest to you. Smooth out any wrinkles.

STEP TWO: If you want, lightly coat the bottoms of the glasses, plates, and other tableware with silicone spray. Silicone spray adds a thin, low-friction coating to surfaces, and it will make the cloth slide more easily. However, it is not necessary.

STEP THREE: Arrange the cups, plates, glasses, and flatware into a place setting on the cloth. For your first attempts, arrange the place setting as close to the edge of the table as possible, and minimize anything that might spill or break easily. As you get the hang of it, move the place setting farther back, add a bowl of fruit, a vase with water and cut flowers, and so on.

STEP FOUR: Stand so that you can pull your arms down and away from the table comfortably. Hold the edge of the tablecloth closest to you with both hands and quickly pull the cloth evenly *down* and *away* from the table. The key is the quick downward motion. Take a bow!

NOTE: It's important that you pull down as you pull away from the table. Trying to pull the cloth straight out without any upward motion is surprisingly difficult, and any upward movement will lift the place setting enough to cause everything to fall. By pulling down, you ensure that the cloth will only move sideways under the place setting.

A classic vaudeville variation of the tablecloth trick is what's sometimes known as the Egg Drop or Eggs in Glasses stunt. A tray or thick sheet of cardboard is placed on top of four tumblers full of water. On top of the tray are balanced four short paper tubes, and on top of each tube is an egg. A sharp, well-placed whack on the side of the tray will cause the paper tubes to fly out of the way, while all four eggs drop precisely into the four glass tumblers below them. We think it's more dramatic, and also easier, to knock the tray out with a broom handle.

MATERIALS

- 4 glass tumblers large enough for the eggs to drop into, almost full of water
- one 12" x 12" piece of corrugated cardboard or a thin tray (cardboard is better to start with)
- two 8½" x 11" sheets of paper (any kind)
- Scotch tape
- 4 eggs (golf balls also work)

TOOLS

- a kitchen counter or sturdy table
- scissors
- a long-handled broom (straw or plastic bristles both work, so long as the bristles are strong and flexible)

HOW TO BUILD IT

The Set Up

STEP ONE: Place the 4 glasses of water near the edge of the table or countertop. Make a square, with each glass 2" to 3" apart.

STEP TWO: Using the scissors, cut a 12" x 12" piece of cardboard (if you aren't using a similar-sized tray). Place the cardboard on top of the glasses so that about 3" of cardboard sticks out over the front edge of the table or counter. Move the glasses closer to the edge to allow this, if necessary. If you are using a tray, make sure that any lip on the tray is up, not down. The tray needs to be able to slide off the glasses without bumping them.

STEP THREE: Using the scissors, cut 4 strips of paper about 4" wide and 5" long. Roll up and tape each strip so that it makes a tube 4" tall and about the diameter of a quarter.

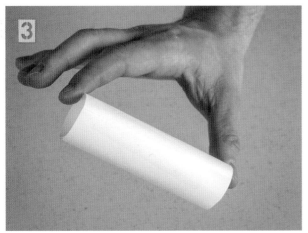

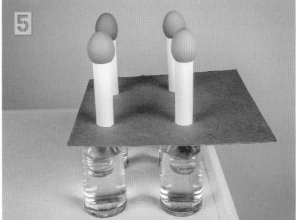

STEP FOUR: Stand each tube on end on the cardboard or tray so that each is directly over the center of a water-filled glass below.

STEP FIVE: Carefully balance 1 egg, on end, on each of the 4 paper tubes. When you're done, it should look like the photograph of step 5.

NOTE: When you're first learning this trick, you might want to use hard-boiled eggs. If things go wrong, you'll have less mess to clean up. Once you get the hang of it, the risk of using raw eggs makes it much more exciting.

The Drop

STEP ONE: Hold the broom a few feet away from the table. Tilt the top of the broom handle toward the table at a slight angle. Then press the bristles into the floor, so that they are bent away from the table.

STEP TWO: Keep holding the broom so that the bristles remain bent and pressed against the floor. At the same time, slide the broom toward the table while pulling back on the broom handle until it's angled slightly away from the table.

STEP THREE: Continue sliding the broom forward until the bend in the bristles is just under the edge of the table. Line up the broom handle with the center of the overhanging 12" x 12" cardboard or tray. Make sure the broom handle is bent back 8" to 12" away from the cardboard.

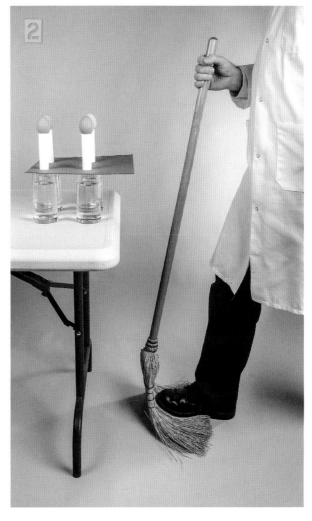

STEP FOUR: Step on the bent broom bristles with one foot to hold them in place securely and pull the broom handle away from the table a few more inches to increase the tension.

STEP FIVE: Without moving your foot, let go of the broom handle so that it snaps out of your hand toward the table. If everything is lined up correctly, the broom handle will hit the protruding edge of the cardboard so that the cardboard shoots out from under the eggs. The broom handle will hit the side of table with a satisfying CLACK, the paper tubes will go flying, and all 4 eggs will drop neatly into the 4 glasses of water. Tada!

THE EXPERIMENT: THE CHAIN REACTION GOLF BALL DROP

Time to take things up a notch. This time you use the force of a falling golf ball to knock out a tube, which lets another ball fall, which knocks out another tube, which lets another ball fall. Unlike the preceding experiments, this trick requires building a Rube Goldberg–like set of tracks and levels. Who's Rube Goldberg? Someone who would enjoy this trick—look him up!

MATERIALS

- three 2½" diameter PVC "Y" joints (a 45-degree Y junction) available at plumbing supply and hardware stores
- duct tape
- 1 paper towel roll (empty)
- boxes, tables, or other stands to raise the height of the device
- 2 pieces of 3" x 1½" cardboard
- 3 toilet paper rolls (empty)
- 1 piece of 4" x 6" cardboard
- 3 golf balls

TOOLS

- Scissors

HOW TO BUILD IT

STEP ONE: Turn each PVC Y joint so that the Y is upside down (one hole points up and two point down). Within each Y, put a strip of duct tape inside so that, when a golf ball drops through, it will roll into the angled side tube rather than dropping straight through.

STEP TWO: Using scissors, cut the paper towel roll in half lengthwise, from end to end, to make 2 half-pipe tracks. These are tracks for the golf balls to roll down.

STEP THREE: Using a combination of boxes, a low coffee table, or chairs, build a small set of 2 steps to place your Ys on. The steps should be about 12" above each other and about 10" deep, with one Y on each step and the third Y at the bottom of the steps.

STEP FOUR: Build the track by connecting the Y joints on each of the 3 steps so that when you drop a golf ball in each, it will pop out the side of the Y and head toward the Y on the step below. Using duct tape, attach one of the half-pipe tracks between the top 2 Y joints: the track should guide the ball out the side of the first Y joint straight into the top of the second Y. Repeat this, attaching the second half-pipe track from the side of the second Y joint to the top of the third one.

Adjust each half-pipe track to make sure that the ball will fall smoothly into the Y joint below. If the end of a half-pipe track is too low or out of alignment, the ball may skip over the next opening completely.

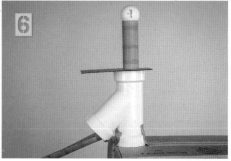

Test your track at this point to make sure the golf balls travel smoothly through the entire course. Adjust or fix any troublesome junctions between the PVC and the cardboard half-pipes, or adjust the steepness of the steps (to slow the ball down), as necessary.

STEP FIVE: Using scissors, cut 2 pieces of 3" x 1½" cardboard. Cut each of the 3 toilet paper tubes once lengthwise. Make each tube smaller by tightening it up until it has a 1½" diameter, and tape the outside of each tube to keep if from unrolling. These smaller-diameter tubes will allow the golf balls to sit more on top of the tubes rather than nestling too far down inside them.

Tape the middle of one of the 3" x 1½" pieces of cardboard onto the end of one of the tubes so that the cardboard makes a base that allows the toilet paper roll to stand up on end. Repeat this with the second piece of cardboard and a second toilet paper roll. These will stand up on top of the bottom 2 Y joints.

STEP SIX: Using scissors, cut a piece of 4" x 6" cardboard. Place this over the hole of the topmost Y

joint, and stand the last toilet paper roll (without a base) on top of it, so it is centered over the hole.

STEP SEVEN: Place the 2 cardboard tubes with bases over the centers of the holes of the second and third Y joints. Place a golf ball on top of each of the 3 toilet paper rolls.

STEP EIGHT: It's showtime. With your hand, sharply hit the 4" x 6" cardboard under the first toilet paper roll, just like the broom did in the Egg Drop. The ball should fall straight down into the first Y joint, travel down the first half-pipe track, and knock out the cardboard tube over the second Y. Then the first ball (followed by the second) should continue into the second Y joint, down the second half-pipe track, and knock out the cardboard tube over the third Y joint. Finally, all 3 golf balls should fall into the third Y joint and come out onto the floor—plunk, plunk, plunk, one after the other.

This is just one crazy contraption you can build that uses the Egg Drop idea as part of a chain reaction. See what else you can come up with!

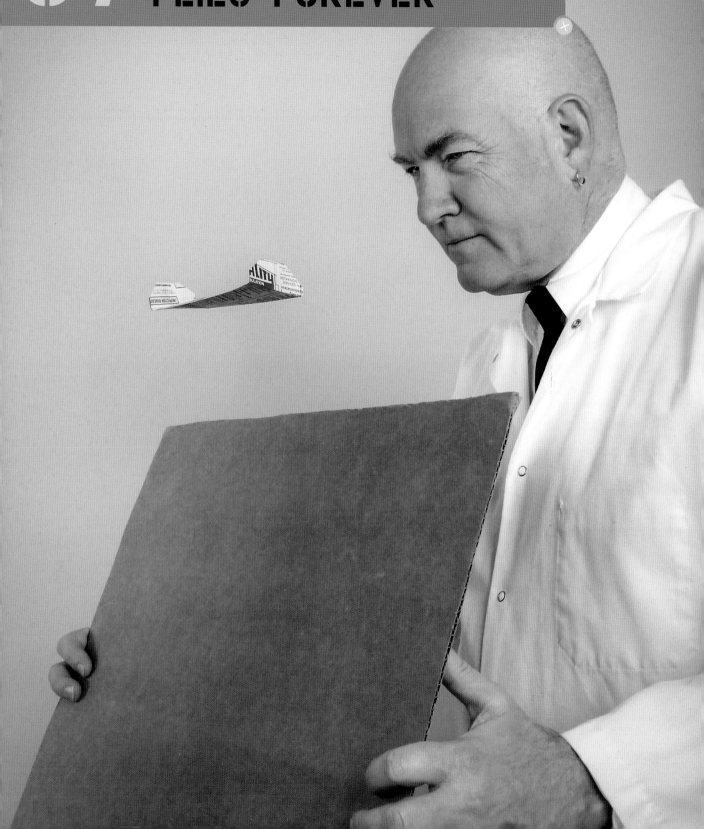

Why isn't everything on the table pulled sideways with the tablecloth or at least knocked down? How can you whack cardboard from under an egg and not shift the egg sideways? And especially, why does this work better the faster you remove the tablecloth or the cardboard, when more force would presumably be more likely to affect the other objects on top?

Here's why.

First, there's *inertia*: an object at rest will stay at rest and an object in motion will remain in motion, unless acted upon by an external force. Newton's First Law of Motion describes this in slightly more complex and precise terms, but that's the gist of it. So, thanks to inertia, the place setting will just sit there unless we push or pull it in some way.

When you pull on the tablecloth, *friction*, the resistance that one surface encounters when sliding over another, comes into play. Friction between the tablecloth and the place setting wants to drag the place setting along for the ride.

So which will win? Will inertia keep the place setting stationary or will the force of friction pull it right over the edge of the table?

If you can overcome the friction of the tablecloth, inertia will keep the table setting in place so it will drop almost straight down.

The first thing that will help you is making this a slippery situation. A slippery tablecloth or some silicone spray reduces the friction and increases the odds of success. With the Egg Drop, though, you can't make it any slipperier, but you can make it fast.

Speed is the next important factor. Pulling quickly reduces the amount of time friction exerts force on the place setting, once again allowing gravity to triumph over friction.

Finally, a heavier place setting has more inertia (it takes more force to move it), so it's more likely to stay put. But if it's too heavy, it'll be hard to overcome the force its mass puts on the tablecloth. For instance, imagine how hard it would be to yank a tablecloth out from under the tire of a bus.

So, ideally, for the tablecloth trick, you want heavier objects (but not as heavy as a bus) so that the inertia (which will keep them in place) is relatively significant compared to the small amount of frictional force (which will pull them sideways) caused by the tablecloth quickly sliding out from under them.

Once the tablecloth is gone, or once you knock the cardboard tubes out from underneath, gravity pulls the objects that remain above straight down. With luck, they land where you want them to: onto the table top, into the water glass, or into your multileveled, PVC Y-joint contraption.

For more ideas, videos, and variations, visit www.eepybird.com/experiments/eggdrop.

How long does a paper airplane stay in the air? Five seconds? Ten seconds? How about (seemingly) forever? The paper airplane in this experiment is super simple to make, and while flying it takes a little while to master, once you do, it will fly farther than any plane you've ever encountered.

HOW DOES IT WORK?

This plane is a simple version of what are known as *walkalong gliders*—it will stay aloft forever so long as you walk along with it. In fact, this design is so different and so effective, when paper airplane guru John Collins first entered one in the "time aloft" competition at a paper airplane festival, they had to change the rules of the competition to disqualify this entire class of plane design.

Theoretically, this plane will fly for as long as you continue to walk with and guide it. Unfortunately, to successfully fly this plane for miles, you'll need a very calm, windless hot summer day. Or, you'll need to walk laps in a gymnasium.

There are lots of designs for walkalong gliders, including some fancy ones that look like real planes— one design even has a set of realistic-looking rotating propellers. However, this design, known as the

Tumblewing, and invented by John Collins, is the quickest and easiest to make.

How does this paper airplane stay aloft? Have you ever seen seabirds hovering steadily over the ocean above a cliff, or a hawk or a vulture soaring over a valley without flapping its wings? If so, then you've witnessed the phenomenon that keeps the walkalong glider aloft. It's called *ridge lift* or *slope lift*.

Slope or ridge lift occurs when air hits a slope and is forced upward. This updraft creates a rising elevator of air directly windward of the cliff. Wind coming off the ocean hitting a hill or a cliff can create nearly continual ridge lift. Soaring birds know this and can spend long periods aloft riding these air currents without flapping their wings.

For this glider, you create slope lift by walking with the plane while holding a large sheet of cardboard.

Ridge Lift, Wave Soaring and the Walkalong Glider

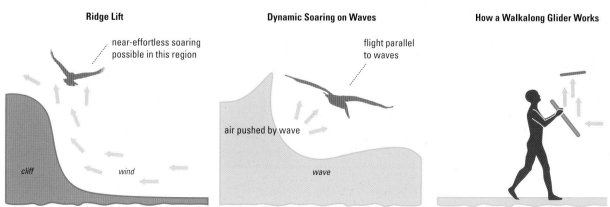

Ridge Lift
near-effortless soaring possible in this region

cliff

wind

Dynamic Soaring on Waves
flight parallel to waves

air pushed by wave

wave

How a Walkalong Glider Works

MATERIALS

- the glider pattern on this page

OR

- the glider pattern from EepyBird.com (that you can download for free)
- 8½" x 11" printer paper (if you're downloading the digital image)
- page from a phone book
- Scotch tape
- 2' x 2' or larger piece of cardboard (at least as big as the top of a large pizza box)

TOOLS

- scissors
- printer (if you're downloading the digital image)

HOW TO BUILD IT

STEP ONE: Photocopy the glider pattern shown below and cut it out. You can also download the pattern from EepyBird.com and print it out.

STEP TWO: Remove a nonessential page from a phone book and lay the pattern on top of it. Phone book paper is just the right weight for this experiment–tissue paper is too flimsy and regular paper is too heavy.

Using 2 small pieces of Scotch tape, tape the ends of the pattern to the phone book paper. Using scissors, cut around the pattern to get a rectangle of phone book paper with the corners cut off, but leave two little points where the tape attaches the pattern to the paper. You'll cut off those points shortly, but for now, you want to keep the pattern and the phone book paper sandwiched together so you can fold them both together along the lines shown on the pattern.

STEP THREE: Fold both ends up 90 degrees, about 1½" in from the end (where the dotted lines are on the pattern).

STEP FOUR: Fold the one edge (what will be the leading edge) down at a 45-degree angle (where the dotted line is along the top of the pattern). Fold the other edge (what will be the trailing edge) up at a 45-degree angle (where the dotted line is along the bottom of the pattern).

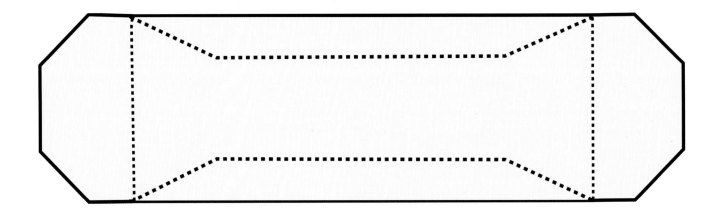

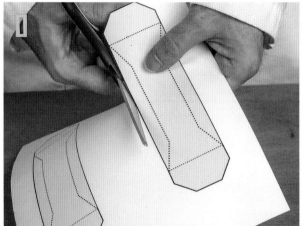

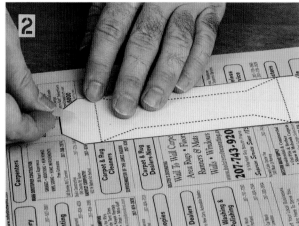

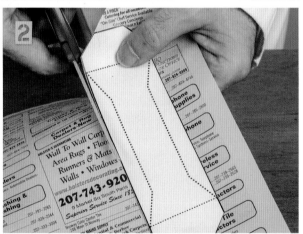

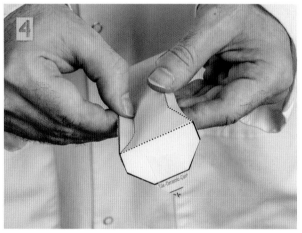

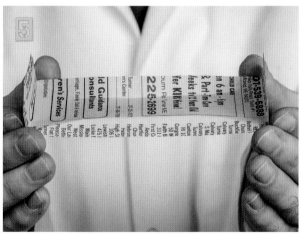

STEP FIVE: Now that the glider is folded, you can cut off the two points where the tape holds the pattern to the phone book paper. Separate the pattern from the phone book paper. You're done with the pattern, and your glider is now complete.

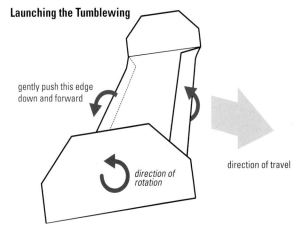

Launching the Tumblewing

gently push this edge
down and forward

direction of rotation

direction of travel

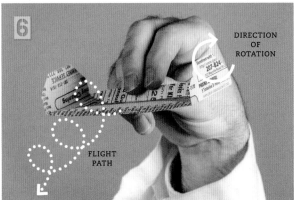

DIRECTION
OF
ROTATION

FLIGHT
PATH

STEP SIX: Hold the center of the Tumblewing's trailing edge (the long, rear angled-up flap) between your thumb and second finger.

With your other hand, hold the large piece of cardboard vertically in front of you at a slight backward tilt of 20 to 30 degrees. The top edge of the cardboard should be at about the height of your shoulders.

STEP SEVEN: Hold the Tumblewing 10" to 12" above and 8" to 10" ahead of the top edge of the 2' x 2' cardboard. Release the Tumblewing with a slight forward and downward push. You want to push the trailing edge that you're holding down and forward so that the Tumblewing itself begins to spin backward, while it drops down and forward.

STEP EIGHT: As soon as you release the glider, begin to walk forward slowly while holding the 2' x 2' cardboard in front of you. This will create an updraft under the Tumblewing that will keep it aloft for as long as you keep walking forward. If you turn left or right, the plane will, too. By walking in a large circle or figure 8, you can keep the plane flying in a relatively small space for a while.

NOTE: It can take practice to get the hang of it. One common mistake is to give in to the temptation to tilt the cardboard at more of an angle to try to "lift" the glider from below. Don't do it. The cardboard needs to be close to vertical to get the necessary slope lift. If your glider is dropping too fast, walk faster, don't tilt the cardboard back. Also, if you find your Tumblewing turning on its own to one side or the other, check to make sure the ends are as straight and vertical as possible. If they angle in or out, they will make the plane veer to one side.

THE SCIENCE

Slope lift is a common phenomenon. We observe it most easily when seagulls hover nearly effortlessly above dunes or cliffs while facing into the wind, or when we see raptors soaring along mountain ridges for miles without flapping.

Over water, pelicans, albatross, and a few other seabirds use slope lift to glide in exactly the same way that we use cardboard to keep the walkalong glider aloft. That is, the moving air doesn't hit a stationary cliff or mountain ridge; it hits the moving slope of the waves. This creates a small mass of rising air in front of each wave, and some seabirds take advantage of this slope lift to glide.

On the coast, where rolling waves come in and hit the shore, pelicans will fly out to a breaking wave and soar parallel to, and right in front of, a breaker, riding the slope lift without needing to expend energy to stay aloft.

The albatross, however, is the true master of this kind of flight. With a wingspan of some 12 feet and weighing up to 24 pounds (about five bags of flour or sugar), the great albatross is one of the largest flying birds in the world. To get and keep that much weight airborne is quite a feat, but once in the air, the albatross can stay aloft and fly for miles without flapping its wings even once. It does this by slope soaring across the open ocean.

As happens near to shore, steady breezes in the open ocean create regions of rising air above the undulating surface of the sea and in front of the waves that ripple without breaking across the surface. The albatross has mastered these drafts and can use them to fly 500 to 600 miles a day low over the surface of the open ocean without ever once needing to flap its wings.

Slope lift can be a powerful and predictable force. Human glider pilots use it to stay aloft as well. Where geological features create long regions of slope lift, pilots can fly their engineless gliders—which can weigh as much as 1,500 pounds—almost indefinitely. As glider pilots learned to master slope soaring in these areas, glider competitions had to discontinue time-aloft contests for fear of pilot exhaustion. Distance competitions for gliders now typically require round-trip courses, and skilled glider pilots have been able to cover as much as a thousand miles of round-trip unpowered flight.

THE WALKALONG GLIDER AND THE TUMBLEWING

Birds have been using slope lift for millions of years, and paper airplanes have been around since the Chinese began to manufacture paper (some twenty-five hundred years ago), but using slope lift for a walkalong paper airplane wasn't invented until the 1950s. Joseph Grant received the first patent for a walkalong glider in 1955, and walkalong glider guru Tyler MacCready patented an improved design in 1992. The walkalong design essentially mimics the flight of a pelican in front of a breaking wave.

As the paddle (your piece of cardboard) moves through the air, it creates an area of continual slope lift in exactly the same way that a moving wave does. The walkalong glider floats on that rising air just like a pelican riding in front of a breaking wave.

A walkalong glider like the Tumblewing is designed to fall steadily forward and down. By adjusting the speed of your walk and the angle of the cardboard, you can create an updraft in front of you that blows the glider up at the same rate that it falls down, so it stays at the same altitude as it moves forward. That's why if you walk too fast, your Tumblewing will fly up and over your cardboard sheet, and if you walk too slowly, it will fall out of the rising air created by the moving slope and tumble to the ground.

The Tumblewing is just a taste of what walkalong gliders can be. Using his innovative designs, Tyler MacCready has flown a walkalong in front of him using just his forehead to create the slope lift. It looks like true levitation.

 For more ideas, videos, and variations, visit www.eepybird.com/experiments/airplaneforever.

Want to magically walk right through a giant spiderweb without it touching you? This fun trick lets you create a video that shows you standing behind a web made of tape that stretches across a doorway. When you walk forward, you appear to melt right through the tape until you're suddenly standing in front of the web!

HOW DOES IT WORK?

Two-dimensional images like videos flatten three-dimensional reality and force our brain to make some guesses about what we're looking at. As we saw with The Face that Follows (page 12), our brain can misinterpret what it sees. In particular, we guess at how big objects in a two-dimensional image are, based on things like how much space they take up in the image and what context they appear in.

Because of this, our brain can be deceived. Peter Jackson's *Lord of the Rings* movies do this a lot. He films in a way that tricks us into thinking that all those adult actors playing hobbits are much smaller than they actually are. This optical illusion is called *forced perspective*.

With the right camera angle and the right alignment of objects, the enormous Taj Mahal can be made to appear small enough for someone to simply reach out and touch the top.

Here, you use the technique of forced perspective to make an image that looks like a spiderweb stretched across a doorway. Actually, it's two separate webs that only look like one.

THE EXPERIMENT: THE SPIDERWEB ILLUSION

This illusion requires a doorway with enough space on one side for the video camera to capture the entire doorway in the image. For the photos in this book, we used a double doorway to make the illusion bigger, but a single doorway will work as well. This project also requires two people: one person builds the web while the other looks through the video camera and guides the web builder.

MATERIALS

- 1 roll of masking tape (blue painter's tape works particularly well)

TOOLS

- a video camera on a tripod
- a ladder or step stool (to reach the top of the door frame)
- 2 small lamps (to adjust the lighting of your video)

HOW TO BUILD IT

STEP ONE: Position the camera on the tripod at an angle to the door and off to the right of the door so that the entire doorway appears in the image. The photo shows a good position for the camera. Once you've got the camera in position, don't move it or adjust the zoom. It must remain fixed in place from now on.

NOTE: It doesn't matter on which side of the doorway you place the camera. However, if you put your camera to the left of the door, reverse the directions listed below (that is, left becomes right, and right becomes left). Further, remember that all the directions we give below are from the point of view of the camera. So when we say "left," we mean left from the perspective of the person behind the camera looking at the door.

STEP TWO: Start making the masking tape web with 3 pieces: 1 connects from the center of the door frame straight down to the floor, and 2 connect from the center of the door frame to the top corners of the door frame, so that it looks like you're making a peace sign.

STEP THREE: Now fill in just the right side and the top of the web, running more strips out from the center of the web to the door frame and then connecting them with short strips of tape going around the web.

HERE'S WHAT THE CAMERA WILL SEE.

HERE'S WHAT YOU'RE ACTUALLY GOING TO BUILD.

STEP FOUR: Now here's the tricky part. You're going to run strips of tape along the wall beyond and behind the door frame so that, from the point of view of the camera, it looks like you're completing the web. This is easiest if there's a wall running back behind the left side of the doorway (as there is in the pictures), but if there isn't, you can position some furniture behind the doorway so that you can connect the tape to the furniture.

Start by attaching a long piece of tape to the left side of the door frame and have one person (the builder) hold it back in the room near the wall (or furniture). The other person (the guide) will look through the camera's viewfinder and tell the builder to move the tape up or down until it looks like it goes right to the center of the web.

Because the far end of the tape is farther away from the camera than the rest of the tape, that end of the tape will look too skinny to the camera. So layer a second (and if necessary, a third) long piece of tape from the same starting point on the door frame to a slightly higher and/or lower end point. This "fattens" the width of the tape on a progressive angle, but in the viewfinder, this will make the tape appear to be the same width for its length. Again, the guide looks through the viewfinder and directs the builder on how to position the tape to achieve this effect.

STEP FIVE: In the same manner, continue to fill in the lower left part of the spiderweb along the wall, with the guide directing the builder to position each piece of tape. As before, each web "strand" will actually be made of multiple strips of tape that fatten the far ends so that they look like the same thickness in the viewfinder.

STEP SIX: Once the web looks right in the viewfinder, use the 2 lamps to adjust the lighting, which should be as even as possible across the entire web. If one part looks significantly brighter or darker than the other, that weakens the illusion. We find it works well to have one light shining on the web in the door frame and one light behind shining on the web that stretches behind the door frame.

STEP SEVEN: Press record! You're ready to have fun with your web illusion and capture it on video to show your friends. Film yourselves walking through it and tossing things back and forth. What else can you think of? Once you've got your video, see if your friends can figure out how you did it!

Forced perspective is a great way to make a toy dinosaur look like it's attacking you or to make you look as tall as a building. How does it work? It's all because a camera has only one "eye," and we have two.

Our two eyes give us slightly different views of the world, which, as mentioned in the Face that Follows, helps us judge depth. Movies in 3-D send each eye a slightly different image, which mimics the way our eyes work in the three-dimensional world, but a regular camera has just a single point of view, and everything it sees might just as well exist on the same flat plane. The "binocular vision" of our eyes can help discern, for example, that a tree is far behind a person, but the two-dimensional image a camera produces can make it look like the tree is growing right out of the person's head.

Most of the time, two-dimensional images work well. Our brains use experience and other cues to judge the relative depth and the relationship of objects to each other, yet when objects line up in particular ways, our brains are fooled and it can look as thought someone's fingertips are touching the top of the Taj Mahal.

To create the forced perspective illusion, we need two important components. First, we need to align separate objects in such a way that they seem to be relating to each other. Second, we need to remove all the clues that contradict the illusion. (For instance, if two objects are lit differently, that can be a giveaway.)

In *The Lord of the Rings*, Peter Jackson put Elijah Wood, who played the small hobbit Frodo, farther away from the camera than Ian McKellen, who played the normal-sized wizard Gandalf. But if that's all he had done, Frodo would have just appeared farther away, so Jackson reinforced the illusion in other ways, such as by placing an oversized mug next to Frodo and a similar but much smaller one next to Gandalf. He also worked hard to make it appear that the actors were looking right at each other, even though they were actually too far apart to do so. Equally important in this kind of illusion are things we don't see, such as the floor. If we saw the big expanse of floor between the Wood and McKellen, we would instantly realize how far apart they were.

Want to try your own version of this trick? Here's a simple one. Record a video of one person pretending to push a box while another *giant* person stands behind watching. All you need are a few boxes and a video camera. Line everything up just right, and what the camera sees is this:

Here's a photo of what you're actually doing, and the real position of each person:

Don't show the floor. That would break the illusion. Experiment with forced perspective and see what else you can come up with!

For more ideas, videos, and variations, visit www.eepybird.com/experiments/spiderweb.

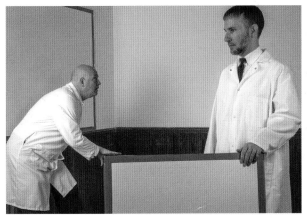

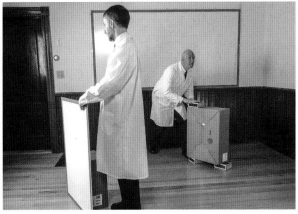

*A WORD ABOUT TOOLS AND SAFETY

A lot of the experiments in Level II and Level III of this book involve power tools. Quite a few involve projectiles. One involves fire and a brazillion flying sparks.

Done carefully, with a modicum of attention, they're all quite safe. But often the same thing that makes them so fun to do can make them a little dangerous if you get careless.

Pretty much every power tool out there can cause injury if not used with proper care. That doesn't mean you shouldn't use power tools, just that you need to have safety on your mind when you do.

The same is true for the experiments in this book, and for the ones you'll inevitably do on your own as you explore these kinds of things yourself. Keep safety in mind. Before you set things in motion, do a final safety check. Ask yourself what might go wrong. If you're heating or burning something, ask yourself if anything could accidentally catch fire, boil over, burn something, or burn someone, and if so, do what you can to minimize that risk. If you're making a projectile, ask yourself what could happen if there's a misfire or your device breaks apart just as it's firing (which early prototypes often do), and be prepared for that to happen. If you're building something you'll ride, make sure you've got an emergency braking method and a plan for what to do if your prototype begins to break apart while you're on it.

In general, always think about what might possibly go wrong before you start, and—before you start—have a plan to deal with those things.

Finally, a word about eye protection. We love the look of the goggles we wear in our videos, but we actually wear them to protect our eyes. You should, too.

The U.S. Occupational Safety and Health Administration estimates that about 1,000 Americans suffer an eye injury *every single day*. That's 30,000 a month and 365,000 every year. Something as little as a small shard of plastic flying up into your eye while you're drilling a hole in a bottle cap could cause a real problem. Don't let that happen. Don't be one of the statistics.

An inexpensive pair of safety goggles can offer pretty good protection in the event that something goes wrong. You can get a pair of shop goggles for a couple of bucks at any hardware store. Pick up a pair or two when you're out buying supplies for your experiments *and use them*.

Besides, they make you look cool.

Lab coats are optional, but of course stylish.

A WORD ABOUT MATERIALS

A bunch of the experiments coming up use lumber and screws. Most of the time, we use inexpensive 1" x 3" lumber that comes in 8' lengths, commonly called strapping. So throughout these experiments, we use the word strapping to refer to 1" x 3" lumber. And don't be deceived by the dimensions. All lumber like this is a bit smaller than advertised, so 1" x 3" means it actually measures ¾" x 2½" when you bring it home. That's to be expected.

To attach pieces of wood, we like to use drywall screws. They're also inexpensive and do a good job of biting into wood without the need for pilot holes to get them started. We start most big projects with a box of 1¼" drywall screws and a box of 2" drywall screws nearby. Whenever an experiment asks for a screw shorter than 1¼", however, use a wood screw.

LEVEL II

TAKING IT UP
A NOTCH*

This is the coolest science experiment ever invented that you can do at home. Everyone should do a Coke and Mentos Geyser at least once in his or her life. And, the next step, doing a ten-bottle fountain display, like a miniature version of the Bellagio fountains in Las Vegas, is unforgettable—and really easy!

HOW DOES IT WORK?

Inside every bottle of soda—or any carbonated beverage, for that matter—is carbon dioxide gas (CO_2), under pressure, waiting to get out. That distinctive hiss you hear whenever you open the bottle is some of that CO_2 escaping into the air. That's why soda, beer, and champagne foam up when you pour them, and why carbonated beverages will continue to bubble for almost an hour after they are opened. Those bubbles are the carbon dioxide continuously escaping from the liquid.

But why, you might wonder, do carbonated beverages react so explosively to Mentos? Microscopically, the surface of every Mentos candy is very bumpy. That bumpy surface—and not the ingredients in the candy— is what makes these geysers explode so fast.

When you drop a handful of Mentos into a glass of soda, the bumpy surface of the candies triggers a reaction in which almost all of the carbon dioxide in the soda escapes all at once. If performed in an open-topped glass, this makes for an amusing, short-lived, messy eruption of foam. However, when the CO_2's only avenue of escape is a narrow hole in the cap of a 2-liter bottle, the release of pressure is quite dramatic. All those bubbles create a stream of soda and carbon dioxide that shoots a jaw-dropping twenty to thirty feet in the air.

We'll start our Coke and Mentos experiments by setting off just one bottle to get an idea of how things work. We could do just that all day—and we have. But eventually, firing up single bottles only whets your appetite for something, well, a little grander. Our first ever multibottle attempt was captured on video, which you can see at EepyBird.com. That first video used ten bottles, just like this experiment, and it made us very happy. Of course, if you're really ambitious, there's no limit to how many you can do. The world record for simultaneous geysers is now over three thousand!

THE EXPERIMENT: THE SINGLE GEYSER

Let's start small and simple—and spectacular! A single geyser is a lot of fun. With this experiment, you can probably stay dry, but be prepared to get wet. It's part of the geyser experience!

MATERIALS

- 1 bottle of room-temperature diet soda (not cold)
- 6 Mentos (just under ½ a roll)
- 1 sheet of 8½" x 11" paper
- Scotch tape
- 1 playing card, business card, or similar small piece of cardstock

WHAT KIND OF SODA?

We've used both name-brand and off-brand colas, and we've used both Diet Coke and Coke Zero. We like using Coke Zero these days.

Most people are surprised to learn that all carbonated beverages work. Both diet and regular soda work beautifully. Even beer and champagne work, but champagne is a bit expensive for these sorts of experiments.

We recommend using diet soda for two reasons. One, we think diet goes just a touch higher. It tends to have a bit more carbonation than regular soda. But more importantly, diet soda has no sugar, so it's not at all sticky to clean up, and without the calories, the soda seems to hold no interest for bugs. We've actually seen bees come over, check out the bottles, turn around, and fly away. Of course, we're not saying diet soda *doesn't* make a mess; it's just that diet soda makes much *less* of a mess.

So you can use pretty much any kind of soda, but diet cola works really well.

HOW TO BUILD IT

STEP ONE: Find a good spot outside, away from power lines and anything else that might object to getting wet. On a fairly level patch of ground or in an empty driveway, set all your materials out. Open the bottle of soda and break open the roll of Mentos. Set aside 6 mints.

STEP TWO: Roll the piece of paper into a tube 8½" long and just a little bit larger in diameter than the roll of Mentos. The tube needs to be just big enough for the mints to fit inside. Tape the outside of the tube to keep it from unrolling.

STEP THREE: Hold the playing card across the bottom of the tube with one hand, and with the other, drop the 6 Mentos into the tube so they land on top of the card.

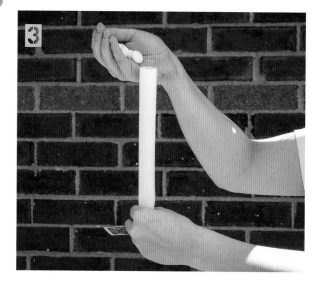

STEP FOUR: Position the card and the tube on top of the bottle. Make sure the tube is lined up exactly with the mouth of the bottle. Make sure the card does not slip, and no Mentos fall out.

STEP FIVE: When everyone's ready, pull out the card while holding the tube in place, positioned over the bottle. Hold the tube just long enough for all the Mentos to fall into the bottle. Once they're in, take a few steps back and enjoy the effect!

USE WARM SODA!

Temperature is the key to great geysers! Use warm (that is, at least room temperature) soda. It's tempting to take a bottle right out of the refrigerator and use it, but if you do, the results will be disappointing. As with most physical or chemical reactions, the warmer things are, the more energy there is in the system, and the better it works.

Some of the highest fountains we've ever set off were the ones we did in New York for *The Late Show with David Letterman*. We were setting up for hours outside on 53rd Street in 90-degree weather, and the results were fantastic. Some of our most disappointing fountains were some of the earliest, when we were experimenting in 40-degree weather

in March and April at EepyBird Headquarters in Maine. Eventually, we realized that warmer is better, but it took us an embarrassingly long time to figure this out!

WARNING!

You may be tempted to try *hot* soda, but this is not necessary and could be dangerous. Room temperature is warm enough. If your soda is cold, and you want to warm it a bit, let it sit in a sink of hot tap water for a few minutes. *Do not heat soda bottles in the microwave or on the stove top!* They can melt and/or explode, leaving you with a huge mess and perhaps personal injury to deal with.

THE EXPERIMENT: THE 10-BOTTLE EXTRAVAGANZA

That single geyser is fun, and setting off more is even better. For this one, you should definitely plan on getting wet.

MATERIALS

- 10 extra bottle caps
- 4 rolls of Mentos mint candies (at least 50 individual candies, but have extra on hand)
- fishing line or thread
- 10 binder clips
- resealable plastic bag
- ten 2-liter bottles of diet soda (but have extra on hand)
- 10 sticky notes
- 1 Lazy Susan

TOOLS

- drill with ¼" and ¹⁄₁₆" bits
- channel lock pliers
- scissors
- pushpin
- safety goggles

HOW TO BUILD IT

The key to the multibottle Coke and Mentos Geysers is this specially made cartridge:

When you're ready for the show, you just remove the binder clip. That will release the Mentos so that they fall into the soda, causing the geyser to erupt out of the hole in the cap. Since the hole in the cap is smaller than the mouth of the bottle, the pressure sends the geyser much higher.

CARBONATION CONSERVATION

Wait as long as possible to open the soda bottles. We don't like to open the bottles until we're just about ready to set them off. That's why we tell you to make your cartridges ahead of time using extra bottle caps. Then, when your cartridges are ready and everything is arranged, open the bottles and load the cartridges. That way the soda doesn't sit around slowly going flat for an hour or two while you're working. Some of the effects we were most looking forward to in our first video completely fizzled because (we think) we opened the bottles too long before we set them off. Don't let this happen to you! Preserve your oomph!

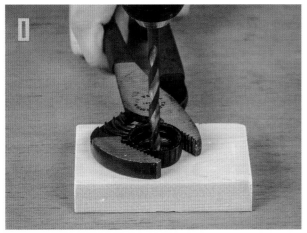

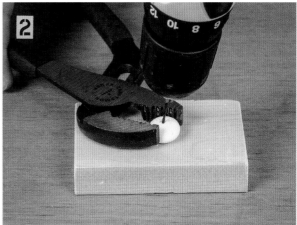

Making the Cartridge

The first and most complicated step is preparing 10 cartridges of 5 Mentos mints each on a fishing line.

Basically, each cartridge consists of a bottle cap with a hole drilled in it, a length of fishing line threaded through the hole, and some Mentos threaded onto the fishing line, which hangs below the cap. We put a binder clip on the fishing line above the hole to keep the Mentos from falling into the soda until we're ready.

Where do you get the "extra" bottle caps to make the cartridges? From previously emptied 2-liter soda bottles, of course, whether they were emptied for your own refreshment or from use as single-bottle practice geysers.

WARNING!

When making the cartridges, *DO NOT hold the bottle cap or the Mentos with your fingers while drilling*. Seriously, it's very dangerous. The holes are trickier to make than it sounds because the caps and the Mentos are small and somewhat slippery. You never want to have your fingers that close to a spinning drill bit. As the instructions say, use pliers, not your fingers. Pliers can be replaced.

STEP ONE: Using the drill, drill a ¼" hole in 7 of the extra bottle caps and a ¹⁄₁₆" hole in 3 of them. Here's the safest method: with one hand, use a pair of channel lock pliers to hold the cap upside down on top of a block of scrap wood. Work the drill with your other hand. Remember: Heed the Warning and don't use your fingers to hold the bottle caps.

A word about hole size: We've tried a bunch of different size holes and ¼" seems to work best. A ½" hole makes for a big, impressive, but short-lived geyser, while the ¼" hole is not as thick but it goes higher and lasts longer. With a 10-bottle fountain, the extra time lets the spectacle build. That said, feel free to use ½" or other size holes. Experiment and create your own combinations and sequences of geysers.

STEP TWO: Drill a ¹⁄₁₆" hole through the center of each of 50 Mentos. Again, *do not hold the Mentos with your fingers* as you drill holes in them. As with the caps, when you drill, use channel lock pliers to hold the Mentos on top of a scrap piece of wood.

You'll also notice that one side of the candy is slightly flat and one side is slightly rounded. Drill into the flatter side, since the drill bit is less likely to go sliding off.

Finally, a little Mentos math: each roll has 14 mints, and rolls are most economical when bought in packs of 6.

Each 6-roll pack has 84 Mentos, so get 2 packs. Then you'll have enough for two 10-bottle fountains, plus breakage and other attempts. If you don't use them all? You know, the mints aren't half bad.

STEP THREE: Using scissors, cut off about a 5" piece of fishing line, thread a single Mentos on the line, and secure it by looping the fishing line back around the Mentos and tying it off above the mint.

(We know "a single Mentos" sounds weird, but the folks who make them—and they should know—tell us that the word *Mentos*, like the word *moose*, is both singular and plural. So "one Mentos" is correct.)

STEP FOUR: Thread 4 more Mentos onto the fishing line. The first Mentos you just tied on will keep the rest of them from sliding off the end.

STEP FIVE: Thread the free end of the fishing line up through the hole you drilled in one of the caps, and attach a binder clip to it (above the top of the cap) to hold everything in place. Your cartridge is done. Repeat Steps Three through Five 9 more times to make 10 cartridges.

STEP SIX: *Protect your cartridges.* We recommend keeping the finished cartridges in a waterproof container, like a resealable plastic bag, until you're ready for the show. Leaving them lying out on a table is just asking for trouble. Someone might knock an open soda bottle over and accidentally soak all your carefully made cartridges,

ruining them. (Just saying. Not that this has ever happened to us . . .)

Preparing Your 10-Bottle Extravaganza

Ideally, arrange your bottles on a solid folding table, picnic table, or similar raised platform. It's much easier than setting them off on the ground—and it looks way better. Besides, if you use diet soda, cleaning up is not bad. Just hose off the area, let it dry, and that's pretty much it. Or, just do what we often do: wait for the next rain to take care of everything, which it does.

STEP ONE: Write each of the numbers 1 through 10 on a sticky note, and stick one of the notes on each of your 10 bottles to label them. This will help you position the bottles and go through your choreography, since you will start with **Bottle 1** and make your way through to finish with **Bottle 10.**

STEP TWO: Seven bottles—**Bottles 1, 2, 3, 7, 8, 9,** and **10**—will use the cartridges with the standard ¼" hole in the cap. Arrange these seven bottles in two rows on the table, as follows:

Put **Bottles 1, 2,** and **3** in a row in front (farthest away from you and closest to where the audience will be) with **Bottle 1** in the center of the row and **Bottles 2** and **3** to the far left and far right, respectively.

Put **Bottles 7, 8, 9,** and **10** in another row close to the back of the table (closer to you and farther away

from where the audience will be), starting with **Bottles 7** and **8** on the left; leave a gap for the Lazy Susan in the middle of the row, then continue with **Bottles 9** and **10** on the right.

Don't open these bottles yet; just get them in place.

STEP THREE: Remove the cap from **Bottle 4**, and using the pushpin, make a small hole in what we call the "shoulder" of the bottle, about halfway between the mouth of the bottle and the top level of the soda. It's important to take the caps off the bottles here to release the pressure before you start poking holes.

Using the drill, carefully drill a ¼" hole through the plastic wall of the bottle where you poked with the pushpin. Start the point of the drill bit in the pushpin hole to help keep the bit in the right spot. Be careful with this step, since the bottle

may flex and the drill bit may start to wander. Don't drill a hole in your hand, just the bottle!

Repeat this process with the pushpin and drill on the opposite side of **Bottle 4** so that there are 2 holes on opposite sides of the shoulder of the bottle.

STEP FOUR: Repeat STEP THREE with **Bottle 5**, first removing the cap, using the pushpin to make small holes, and using the drill to carefully make two ¼" holes on opposite sides of the shoulder of **Bottle 5**.

STEP FIVE: Place **Bottles 4** and **5** in a row across the center of the table, with **Bottle 4** on the left side of center and **Bottle 5** on the right side of center.

STEP SIX: Open **Bottle 6**, and, using the pushpin, make 3 holes in the shoulder, equally spaced around the circle of the shoulder and all at the same height, about halfway between the neck of the bottle and the top level of the soda. Using the drill, carefully drill a ¼" hole on each of the 3 spots where you made a starter hole with the pushpin.

Place **Bottle 6** on the Lazy Susan in the middle of the back row between **Bottles 8** and **9**.

STEP SEVEN: Put on your safety goggles; from here on, you will begin arming the bottles with the cartridges (but not firing them yet), and you want to protect yourself in case one of them goes off accidentally.

STEP EIGHT: Arm the bottles with Mentos cartridges! One at a time, open **Bottles 1**, **2**, **3**, **7**, **8**, **9**, and **10**, then replace each cap with a cartridge with a ¼" hole in the cap. Screw these on *very gently*, but tightly, making sure the Mentos don't touch the soda and the binder clip doesn't get knocked off by mistake. Once these Mentos touch the soda, things start to happen—see the Warning about accidents! Then, as you arm the remaining bottles, be careful not to bump any of the already loaded bottles.

Once you have screwed on these first 7 cartridges, screw the remaining 3 cartridges with ⅟₁₆" holes in their caps onto **Bottles 4**, **5**, and **6**. For these 3 bottles, most of the soda will spray from the shoulder holes. The ⅟₁₆" holes in the cartridge caps are really just there to accommodate the fishing line.

WARNING!

If you accidentally set off one of the bottles before you're ready to fire the whole fountain display, *don't panic*. Just pick up the gushing bottle, point it away from the other bottles, as well as away from all people and animals, and carry it off to the side. Let it finish erupting where it won't do any harm. This preserves the rest of your set up. Then, if you have extras, replace that bottle, or simply proceed without it.

BOTTLE 6: TOP VIEW

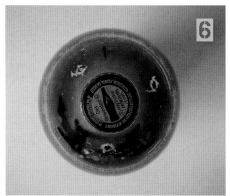

The Firing Sequence

When you're ready to fire, make sure your audience is well out of range and secure any animals, like dogs, so they don't get wet or start running around and knocking over the bottles. Also, make sure people have plenty of room to move in case a bottle tips over while it's firing and starts jetting out toward your spectators.

The Coke and Mentos Geysers here are fired in 5 stages, one immediately after the next. And this is where those sticky notes with the labels 1 to 10 on the bottles will be useful: you start at 1 and make your way to 10.

STEP ONE: Start the show with **Bottle 1**, which is in the front center. Squeeze the butterfly wings of the binder clip, releasing the Mentos into the soda, and step back.

STEP TWO: When the fountain from **Bottle 1** is past its peak and has come down to about 6' high, fire **Bottles 2** and **3** at the same time (or one immediately after the other, as you prefer).

STEP THREE: When the fountains from **Bottles 2** and **3** are past their peak, and down to about 6' high, fire **Bottles 4** and **5** at the same time.

STEP FOUR: A little more quickly, just as the geysers from **Bottles 4** and **5** *hit their peak,* fire **Bottle 6** and *immediately* kneel down (so you will be below the spray). As **Bottle 6** erupts, slowly turn the Lazy Susan under it so that its 3-stream fountain rotates as it sprays.

STEP FIVE: As soon as **Bottle 6** begins to fade, stand up and fire **Bottles 7**, **8**, **9**, and **10**, in that order one immediately after the other.

STEP SIX: Take a step back, watch the last four geysers, and admire the end of the spectacle. When the final geyser has finished, face your audience and throw both your arms up in the air! You did it!

Take pictures. Take videos. Share them with your friends and share them with us on our website EepyBird.com. We want to see them!

VARIATION: SIMULTANEOUS FIRING

Here is one way to release a group of geysers at exactly the same time—while also staying dry, or at least relatively dry. (If you want to stay completely dry, you'd have to skip the awesome rotating fountain effect created by **Bottle 6** on the Lazy Susan, and what fun is that?) All you need is a friend, binder clips, and a long piece of string. In the 10-Bottle Extravaganza, use this as an alternate technique for firing the final back row of bottles, **Bottles 7**, **8**, **9**, and **10**, which look great when set off all at once. To do this, tie 4 binder clips about 2' apart on a single string, with about 4' of additional string on both ends. When your bottles are in place, but before they are armed, carefully replace the binder clips on the cartridges for **Bottles 7**, **8**, **9**, and **10** with the binder clips on the string. Arm the bottles like normal, and when you're ready to set them off, you and your friend each take one end of the string and pull sharply up at the same time. This will pull off the binder clips and all 4 bottles will fire simultaneously.

The process that makes these ordinary bottles of soda erupt into such magnificent geysers is called *nucleation*. Every commercial soda bottle has carbon dioxide (CO_2) forced into it under pressure with a *carbonator*, which is a device similar to the air compressor at gas stations used to fill car tires, except that instead of compressed air, it uses compressed carbon dioxide.

When you open the soda bottle, you release the pressure inside. The CO_2, which has been dissolved into the soda at the bottling plant, now begins to separate itself. Escaping carbon dioxide is what makes the bubbles form when you open the bottle. If you leave a bottle open for an hour or two, it will continue to bubble quietly until all of the carbon dioxide has escaped. When this happens, what you have is flat soda—or flavored sugar water without any carbonation (or, with diet soda, flavored sugar-free water).

Once the pressure in the bottle is released, the CO_2 molecules escape by gathering together into tiny bubbles, which are strong enough to break free of the liquid and reach the air. Water molecules like to stick together, however, and they make it difficult for the carbon dioxide molecules, individually, to break free and rise to the top. First, the carbon dioxide molecules have to gather into larger and larger clusters. When they've gathered together into a large enough group, they can break apart from the water molecules and float up.

The carbon dioxide molecules (like all molecules dissolved under pressure in liquid) first gather at what are called *nucleation sites*. Nucleation sites occur wherever two *phases* of matter (that is, a solid, liquid, or gas) are in contact with each other. So, with soda, the main nucleation sites tend to be where the liquid soda meets the solid sides of the bottle or glass. However, nucleation also occurs where the soda meets the air, and within the liquid it occurs around every growing carbon dioxide gas bubble where CO_2 gas and the liquid soda are in contact.

For this reason, the inside surface of soda bottles is purposefully smooth. A smooth surface limits the space available for carbon dioxide molecules to engage in nucleation. There are very few nucleation sites for the carbon dioxide to start grabbing onto. If the inside of a soda bottle was rough like sandpaper, this would create much more surface area and exponentially more nucleation sites for the carbon dioxide.

In other words, rough surfaces are good for nucleation. If a glass is cracked or etched on the inside, you'll notice bubbles of CO_2 clustering on these rougher areas or spots. Put your fingertip into a glass of soda, and you'll see bubbles of CO_2 form on the tiny ridges of your fingerprint.

The rougher the surface, the more nucleation sites there are, and the faster the CO_2 molecules will group

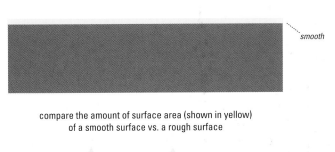

compare the amount of surface area (shown in yellow) of a smooth surface vs. a rough surface

the bumpy surface of a Mentos has lots of surface area for nucleation

together and break free. Typically, soda drinkers want to preserve the CO_2 in the soda for as long as possible; after all, they put the CO_2 in so we could enjoy a sweet carbonated drink. But in this experiment, we don't want to drink the soda. In fact, we want the CO_2 to come out as quickly as possible, perhaps, say, even explosively. So here we want to combine the soda with a solid that has a very rough surface down at the microscopic level.

It turns out that, under a microscope, Mentos mints are extraordinarily rough. We went to the candy factory in the Netherlands that produces all the Mentos for Europe and North America, and we got to see how they're made. One of the things we learned was that each Mentos candy is finished by spraying it with up to 45 layers of a fine mist of liquid sugar.

That spray gives Mentos hundreds of thousands of tiny little bumps and pits made out of hardened liquid sugar. For CO_2, this turns out to be a nucleation-site bonanza.

In fact, Mentos have so many microscopic bumps that when five or six are dropped into a bottle of soda, all of the CO_2 separates from the solution in a matter of moments and forms into large bubbles that all race to escape out of the top of the bottle.

And *that's* why Mentos make soda explode so beautifully.

For more ideas, videos, and variations, visit www.eepybird.com/experiments/cokementos.

THE SEARCH FOR THE BIGGEST GEYSER

When we first started trying this, we wondered if we could make bigger geysers by using more Mentos. What if you drop twenty Mentos into one bottle? What if you used more soda? What if you dropped a thousand Mentos into a swimming pool full of soda?

We found that five to six Mentos seems to provide enough nucleation sites to get all the CO_2 out of a 2-liter bottle of soda almost instantly, so adding more doesn't seem to make the reaction noticeably stronger. Unfortunately, more Mentos won't give you a bigger geyser.

We also found that in every container of soda, no matter how giant, the overall CO_2 pressure in the soda is still the same. So while a bigger bottle can make the geyser last longer (since it has more fizz to release), without more pressure in the system, the geyser won't go higher.

And this led us to the answer for creating the biggest geysers yet (beyond ensuring that the liquid is warm instead of cold, of course): get more pressure. That's where champagne is amazing. There's so much CO_2 in champagne under so much pressure, we've gotten champagne geysers 50 percent higher than anything we've ever seen with soda!

10 FIRE WIRE

Ordinary steel wool is usually used for mundane tasks like scrubbing pots and pans or sanding down woodworking projects. But light it on fire and swing it around your head on a rope, and it makes a dramatic display of homemade fireworks. It's a stunning spinning shower of sparks you and your friends won't soon forget.

Known variously as "fire wire," "steel wool fireworks," "sparkle poi," or "steel wool poi," this firework is surprisingly easy to make. Don't be put off by its simplicity; it makes for a breathtaking display.

However, please heed the warnings: you are playing with fire!

HOW DOES IT WORK?

Yep. Steel burns.

But at these temperatures, it only burns when it's in tiny thread-like strands. Here's why:

Fire is a chemical process in which a fuel (typically something like wood or coal) combines rapidly with oxygen, releases heat, and creates new molecules made up of atoms from both the fuel and the oxygen. This chemical process is called *rapid oxidation*. Although what you're burning in this experiment is called "steel" wool, this particular kind of steel is made mostly of iron, and iron molecules like to combine with oxygen molecules. When iron combines with oxygen slowly, it makes rust. Those reddish brown bits of rust you see on old cars are, in fact, made up of molecules of iron that have combined with oxygen over the years to become iron oxide.

When iron combines with oxygen under the right conditions, it combines quickly and it creates fire. So, yes, iron can burn, and Fire Wire creates just those conditions.

THE EXPERIMENT: FIRE WIRE

MATERIALS

- 1 package of #0, #00, #000, or #0000 steel wool
- 1 wire whisk (preferably with a metal loop at the end of the handle)
- 4-6' of rope or cord
- duct tape (if your whisk doesn't have a metal loop at the end of the handle)

TOOLS

- safety goggles
- gloves
- lighter or matches
- fire extinguisher

WARNING!

As with every project in this book, if you closely follow the instructions and adhere to the safety precautions, you have nothing to worry about. But before you begin this experiment, take a moment to recognize the high (and hot!) risk that is introduced if you stray from the instructions or ignore the WARNING boxes. Note that this experiment should not be done unsupervised by children, and that it's essential to make sure you have a large radius of clear space (70 feet) around you before you set off your fireworks. Have a fire extinguisher nearby just in case. Sometimes maximum fun has hazardous potential—better safe than sorry!

HOW TO BUILD IT

STEP ONE: Prepare the steel wool. Steel wool comes in individual pads that are actually rolls of thin steel wire mesh. Carefully unroll 3 to 4 steel wool pads until you have what looks like 3 to 4 steel filament scarves.

STEP TWO: Gently pull and tease apart the unrolled steel mesh to separate the steel membranes from each other as much as possible without breaking the mesh.

For maximum sparking effect, you want to create as much air space between the steel strands as possible, but you also want the strands themselves unbroken.

STEP THREE: A wire whisk makes a perfect premade metal cage to hold your steel wool. First, attach the rope to the whisk. If your wire whisk has a metal loop at the end of the handle (as many do), loop the end of the rope through and tie it off securely.

If your whisk doesn't have a metal loop at the end, tie the end of the rope firmly around a couple of the wires of the whisk, then run the rope along the whisk handle, and wrap it with duct tape to secure the rope to the handle.

STEP FOUR: Once the rope is safely attached (see the Warning!), gently stuff in as much of the steel wool as will fit loosely inside the whisk. (If you have a small whisk, you won't use all 3-4 pads.)

Preparing for Your Fireworks

For all its dramatic flying sparking magnificence, the Fire Wire is quite safe so long as you follow a few simple precautions before you start. Take extra care when preparing the area where you will light the fireworks and you will have a great time.

Note that the key to this dramatic display is from whipping the burning steel wool through the air. This increases the oxygen available for combustion and makes the fire burn more intensely. However, as a result, this sends burning steel strands flying off in all directions, so you need a wide area free of any other flammable material, and all onlookers need to

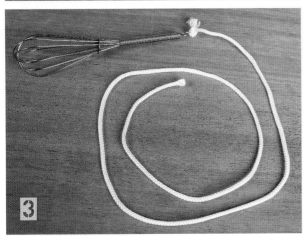

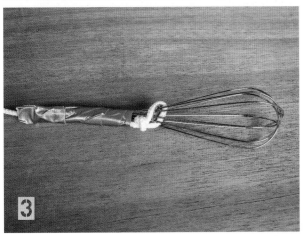

keep their distance. As the burning strands fall to the ground, they tend to burn out quickly. But they do NOT always all burn out *immediately*, so be extremely careful.

SAFETY STEP ONE: Use a truly safe location. "Safe" here means any wide area with surfaces that aren't flammable. Asphalt streets or parking lots (empty of parked cars), cement or asphalt tennis courts, playground blacktops, sand dunes, and beaches are all good places to do this. NEVER attempt this experiment indoors.

SAFETY STEP TWO: When you're wielding the Fire Wire, wear safety goggles and gloves. In addition, cover your skin and hair to protect them from sparks: wear long sleeves and long pants, and a cap or a hooded sweatshirt pulled over your head. Don't have hairspray or gel in your hair—those can be highly flammable. Don't wear lightweight synthetic fabrics like polyester or nylon that burn easily; wear fabrics made of thicker fibers, like cotton.

SAFETY STEP THREE: Clear the area of any potential fire hazards or of anything that might be hurt or burned by embers. Don't underestimate how far the sparks might fly: it could be 50, 60, or 70 feet in a circle. Move personal belongings well clear; sweep away any stray leaves that may have gathered by a fence.

WARNING!

THIS IS FIRE. FIRE IS DANGEROUS.

Fire can be easy to start and hard to stop. Follow every precaution when preparing the area for your fireworks. This experiment generates thousands of sparks that fly in all directions, and it literally takes only one spark to start a fire. If any sparks land on dry grass, leaves, or some other easily ignitable material, they can start a fire that can very quickly burn out of control and turn into a raging inferno. That's no fun. Bring a fire extinguisher, just in case.

SAFETY STEP FOUR: Choose a dramatic time of day. Like most fireworks, Fire Wire is most impressive at dusk or after dark. In daylight, it's hard to see. So, prepare your area in the daylight, but wait until the sun goes down for the show.

SAFETY STEP FIVE: After your Fire Wire display is done, scan the area for any sparks that haven't gone out, and then stay in the area for at least 5 to 10 minutes to make sure no sparks are still smoldering. Use your shoes or a spray bottle filled with water to douse any sparks that don't go out on their own. Never leave until you are 100 percent sure nothing is still burning, and have a fire extinguisher with you, just in case.

Setting Off the Fire Wire

STEP ONE: Put on your safety goggles and gloves. Ignite the steel wool with the lighter or matches. Note that the steel will not become engulfed in flames; rather, individual strands will begin to burn slowly.

STEP TWO: With your gloved hand, grab the end of the rope on the whisk. Then, starting slowly, twirl the wire around in a circle like a cowboy would a lariat. Build up speed until you're spinning the steel wool pretty much as fast as you comfortably can. You'll see some crazy fireworks. Each impressive display should last about 20 to 25 seconds. (You did make more than one Fire Wire, right?)

Also, there are essentially two ways to spin: either in a horizontal plane over your head or in a vertical plane, like a propeller, next to your body. A vertical plane is a little safer because that method sends sparks arcing basically in a line—directly in front and in back of you. Spinning the Fire Wire horizontally over your head sends sparks in a circle everywhere.

THE SCIENCE

Fire requires three things: 1) fuel, 2) oxygen, and 3) heat. Unless all three are present, you won't get fire.

In general, iron doesn't burn because it rarely gets hot enough, and even when it's hot enough, it needs lots of oxygen to get the fire going.

Iron is a good *conductor*, which means that it is a material that transmits energy easily and efficiently. Metals like steel, iron, and copper are excellent conductors.

If you hold the tip of a steel nail over a flame, the heat from the flame will be conducted through the entire nail in pretty short order, quickly making it too hot to hold. This is why you need to use pot holders when cooking with a cast-iron skillet; the iron conducts the heat throughout the entire pan, including the handle, even though the handle isn't the part of the pan on the stove's burner.

Not every material is a good conductor. Take wood, for example. A wooden stick burns easily, but one end can catch fire and you can (usually) still hold the other end comfortably without burning your hand. Unlike the nail or the cast-iron skillet, the heat doesn't travel through the wood very well.

Oddly enough, the excellent conductivity of metal is also what makes metal hard to burn. Consider the cast-iron skillet. In order to burn, it must get extremely hot. The burner on a stove can reach the necessary temperature, but the skillet quickly conducts the heat away from the source (the burner) and spreads or dissipates the heat throughout the entire piece of metal. This usually keeps any one spot—even the spot right over the flame—from reaching the temperature needed to burn. In addition, even if one spot on the skillet does reach a high enough temperature, that spot won't get enough oxygen to ignite. The metal is too thick. Instead of oxygen, the hot iron molecules are mostly next to cooler iron molecules. Without nearby oxygen to combine with, the hot iron can't burn.

Steel wool, however—especially steel wool that you've unrolled and teased out into a loose steel mesh—is different. First, the thin steel strands of the mesh are too thin to conduct heat efficiently throughout the whole steel wool pad. The strands get hot faster and stay hot longer because any heated spot is in contact with only a few other steel molecules in that strand, so it cannot diffuse the heat very effectively to the rest of the steel wool pad.

In addition, the thin metal mesh has a greater surface area than a skillet, so many more iron molecules have access to and can combine with the oxygen molecules in the air.

If you left the steel wool as a tight pad, and didn't pull it into a looser mesh, far less oxygen would come into contact with the metal. This decreases the steel's

FIRE WIRE WITH TEASED-OUT STEEL WOOL.

ability to burn and makes for a less-impressive display. That's why you tease out the steel wool pad till it's a loose bundle.

When you light the steel wool, at first it gets enough heat and enough oxygen to at least begin to smolder, if not burn. Then when you swing the smoldering steel wool rapidly through the air, you drastically increase the amount of oxygen getting to the steel wool, so it burns much more fiercely. It's the same thing that happens when you blow on a campfire to make it burn more intensely

As the iron burns, thousands of tiny glowing sparks fly off the steel wool pad. What's happening is that as the fire burns, the iron and the oxygen are combining to form iron oxide. Iron oxide, or rust, is a more stable molecule, but it's also a far weaker material than iron alone—which is self-evident on any steel object that's rusted through. Thus, as the spinning iron burns and turns to iron oxide, tiny, still-glowing pieces of iron oxide break off and fly into the air, creating the impressive sparking Fire Wire show.

For more ideas, videos, and variations, visit www.eepybird.com/experiments/firewire.

SURPRISE: BURNING MAKES IRON HEAVIER

Surprisingly, burned steel wool is heavier *after* it has been burned. Why? Fire is a chemical reaction in which the fuel combines with oxygen, giving off heat and light. As we've mentioned, when steel wool burns, iron molecules combine with oxygen molecules to form iron oxide, and iron oxide molecules, which contain both iron and oxygen atoms, are heavier than iron atoms alone.

Which is heavier? A molecule of iron (Fe) or a molecule of iron oxide (FeO)?

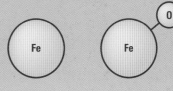

This experiment uses one of the world's simplest motors. All you need is a battery, a piece of copper wire, and a small neodymium magnet. Amazingly, with just those three things, you can build a motor that spins really fast.

It's fast and easy, but it only has a minimal amount of power, so it's tricky to get it to do more than spin by itself. But just like an angel chime, which harnesses the slight power of hot air rising from burning candles, you can build an electric chime driven by this magnet motor.

HOW DOES IT WORK?

This simple device, is a homopolar motor, and like almost all motors, it rotates because of the surprising relationship between electricity and magnetism.

The ancient Greeks, Indians, and Chinese all observed magnetism at least two thousand years ago, noticing that certain stones exerted an invisible force on objects made of iron. It was about a thousand years ago that people found the first important practical use for magnetism: the compass. A floating needle made of magnetic material would always rotate to point north, which allowed travelers to navigate without the stars.

Ancient people also knew something of electricity, both from the shocks you could get from electric fish and from the static electricity generated by rubbing against things like fur. It wasn't until much later that scientists finally found a use for this unexplained force beyond the fun of scuffing your feet on the rug and then getting a little shock when you touched a metal door handle.

As we'll see, it turns out that these two invisible forces are related. This simple motor starts by turning a chemical reaction inside the battery into electricity. That electricity, flowing through the copper wire, generates a magnetic field. And that magnetic field pushes against the magnetic field of the neodymium magnet to generate motion.

The homopolar motor is one of the most elegant illustrations of how electricity and magnetism are connected.

THE EXPERIMENT: THE MAGNET MOTOR CHIMES

First you'll build a motor using nothing but wire, a battery, and magnets. Then you'll build the chimes, with your simple motor sitting right in the middle.

MATERIALS

Magnet Motor

- ▪ 1 length of solid 14-gauge copper wire, about 13"
- ▪ 1 AA alkaline battery
- ▪ 1–2 small cylindrical neodymium magnets, approximately ⅜" diameter, ⅛" thick
- ▪ one 9" length of 3/4" wooden dowel
- ▪ one 3½" x 3½" block of ¾"-thick wood
- ▪ one 1¼" screw
- ▪ double-sided tape

Chimes

- ▪ 1 piece of 4" x 4" cardstock or lightweight cardboard
- ▪ quarter or other small round object
- ▪ fishing line
- ▪ Scotch tape
- ▪ 4 wooden matches
- ▪ 4 water glasses (must be made of glass)
- ▪ water

TOOLS

- ▪ wire cutters
- ▪ needle-nose pliers (many have wire cutters built in)
- ▪ a short length of 1"-diameter dowel or tube, like a ¾" PVC pipe, for bending the wire
- ▪ handsaw or chop saw
- ▪ screw gun or screwdriver
- ▪ scissors
- ▪ ruler

A NOTE ABOUT MATERIALS

When making your motor, keep a few things in mind. Any size battery will work for this, but AA is a good shape. However, this experiment does use up the battery's power fairly quickly, so you might want to have more than one battery on hand.

That said, do NOT use rechargeable batteries. Rechargeable batteries, like lithium or nickel-cadmium, could overheat and become dangerous. Use only regular alkaline batteries.

Never heard of neodymium magnets? You can find them online at science supply sites or on Amazon. If they're strong, you'll just need one, but if they're a bit weaker, you may want more. They're usually about ⅜–½" in diameter and ⅛" high, and are available in packs of 10 for around five dollars.

Finally, use solid copper wire, not braided copper wire, so that the wire will bend and hold its shape. Gauge 14 wire is thin enough to bend easily without being too flimsy.

WARNING!

Don't let small children play with neodymium magnets. They are not toys. Neodymium magnets are small but extremely powerful. In particular, they're very dangerous if swallowed. Two of them can literally stick to each other while they're in different parts of the intestines, requiring life-saving surgery to remove them. Don't put them in your mouth or leave them out where a small child might do so. These magnets are amazing, but they deserve respect.

HOW TO BUILD IT

The Magnet Motor

STEP ONE: Use the wire cutters to cut a 13" length of the copper wire. Then use the pliers to bend one end of the wire, making a 90-degree angle about ¼" from the end of the wire, to get an L-shape, as shown in the photo.

STEP TWO: Bend the rest of the wire into a spiral by wrapping it around the short piece of tube or dowel, as seen in the picture.

STEP THREE: Place one or two of the magnets in a stack underneath the flat, negatively-charged end of the battery. They will stick themselves to the battery magnetically.

STEP FOUR: Loop the coil of wire over the battery with the 90-degree bend touching the center of the positive end of the battery.

STEP FIVE: Stretch or compress the coil so that the bottom of the coil is right next to the magnets, just touching them and nothing else.

STEP SIX: Let it spin! It may take some experimentation with the coil to get the right shape so it will stay balanced around the battery. What happens if you turn the magnets upside down? Try it and see.

Stop the motor by removing the copper from the battery.

The Base

STEP ONE: With the saw, cut the dowel to approximately 9", and cut the block of wood into a 3½" x 3½" square. It's important to make the dowel cuts as straight across as possible.

STEP TWO: Measure and mark the center of the board. An easy way to do this is by drawing diagonal lines connecting the opposite corners of the square; where they intersect is the center. Using the screw gun, attach the dowel to the center of the block with the screw, as shown in the photo.

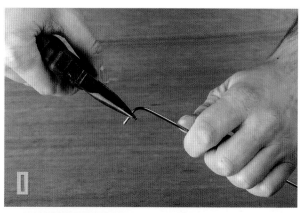

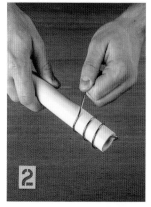

WARNING!

If you let your magnet motor run for a little while, it will get quite hot. Be careful. Don't leave it running too long, and don't leave it running unattended.

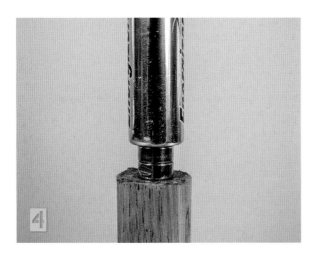

TIP: To make attaching the block to the dowel easier, first drive the screw into the block of wood until the tip has just started to come through the other side. Then push the end of the dowel onto the tip of the screw and continue turning the screw until it is tight.

STEP THREE: Put a small piece of double-sided tape on the upright end of the dowel. That tape will help keep your magnet motor in place.

STEP FOUR: Put the magnets and battery of your motor on top of the dowel, with the magnets stuck to the tape.

The Carousel

STEP ONE: Using scissors, cut the cardstock into a 4" x 4" square.

STEP TWO: Cut a round, approximately 1" hole in the center of the square.

As above, to find the center, mark the diagonals of the square; where they intersect is the center. Then put a quarter or similar small round object over the center and trace around it. If you eyeball this carefully, it should be accurate enough.

STEP THREE: In each corner of the square, using the scissors, make a small cut about ¼" long along the diagonal line.

STEP FOUR: Cut four 5" pieces of fishing line.

STEP FIVE: Here's the tricky part: attach the square piece of cardstock to the magnet coil using the fishing line and Scotch tape. When you're done, the cardstock square will hang horizontally, parallel to the ground, below the coil. First, use the Scotch tape to attach the four 5" pieces of fishing line to the coil, spacing them evenly around. Then slide the other ends of the fishing line into the small cuts in the corners of the cardstock square. Hold the coil up, and adjust the lengths of fishing line until the cardstock square hangs as level as possible. When you've got it adjusted, use the Scotch

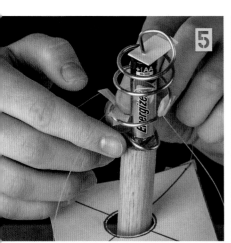

tape to secure the ends of the fishing lines at each corner of the cardstock. Trim off any extra fishing line.

STEP SIX: Cut four 3" pieces of fishing line and tape a wooden match onto the end of each. These will be the mallets that will hit your water glass chimes.

STEP SEVEN: Tape these 4 mallets to the 4 corners of the square so that they hang down below.

STEP EIGHT: Put varying amounts of water in each of the 4 water glasses. The differing water levels will create different tones. (See The Mechanical Water Xylophone, page 180, for more on "tuning" water glass notes.)

STEP NINE: Now, put it all together. Place the entire coil/fishing line/cardstock/matches apparatus over the battery-and-magnet motor on top of the dowel. Everything is perched on top of the dowel. Position the point of the coil on top of the motor and check to make sure that the carousel can spin.

STEP TEN: Position a glass at each corner of the wood base, so that the matches will just glance off them as they go by.

STEP ELEVEN: Let it spin! Music!

REMEMBER: Your magnet motor will get hot. Be careful when you take it apart. Don't leave it spinning for too long, and don't leave it spinning unattended.

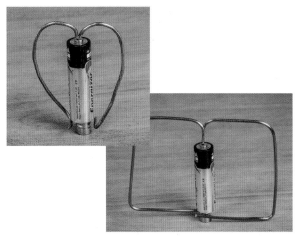

IDEAS TO TAKE IT FURTHER

You can make a lot of different shapes of wire that will spin on top of the battery and magnets. Some of them are difficult to get balanced perfectly, so they tend to fall off after a few seconds, but experiment and see what other shapes you can find.

You can also change what part of the motor spins. Take a look at this:

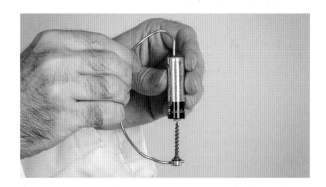

In this setup, the wire doesn't move. It's the screw and the magnet that will spin.

In the early 1800s, a Danish scientist named Hans Christian Ørsted noticed that when a compass got close to a wire carrying an electric current, the needle on the compass moved. Clearly, these two invisible forces must be related somehow.

A few years later, Michael Faraday took it one crucial step further, using the relationship between electricity and magnetism to build the first electric motor. It was a simple motor similar to the one you just built.

It turns out that an electric current moving through a wire generates a magnetic field. Similarly, a magnet moving near a wire generates an electric current in that wire. Faraday used this relationship to build both motors (which turn when you run a current of electricity through them) and generators (which create electricity when you turn them). While we now build them bigger and better, all the motors around your house use the same principles.

You can see Faraday's discovery of the connection between electricity and magnetism in action in a very simple generator that uses only magnets, a coil of wire, and a wheel. The magnets are attached around the edge of the wheel that, when spun, moves the magnets past the coil of wire. As each magnet goes past the wire, the magnetic field creates a little bit of electrical current in the coil. If you connect the wire to a small lightbulb, and then spin the wheel, the lightbulb will light up. The faster you spin this kind of wheel, the more current you get. The more current you get, the brighter the lightbulb glows. It's that simple.

The generator in the illustration is just like the homopolar motor—it's also just magnets and wire—but in reverse. This simple generator uses movement and magnets to generate current. The homopolar motor uses current and magnets to generate movement.

In the homopolar motor, when the current flows from the top of the battery, through the wire, and down to the bottom of the battery, the magnetic field generated by that current hits the magnetic field created by the magnets. Those two fields interact and generate a small amount of force—a very small push, like the way one magnet pushes off another magnet.

That force, called the Lorentz force, causes the wire to spin.

There are equations that describe the Lorentz force and make it possible to predict exactly how these electrical and magnetic forces interact. The equations can be used, for example, to predict which direction a homopolar motor will rotate based on the direction of the current and the polarity of the magnetic field.

When you made your motor, did you try flipping the magnets upside down? If so, then you know that by changing the orientation of the magnetic field, the homopolar motor spins the other way. That's the Lorentz force, the relationship between electricity and magnetism, in action.

Why is it called a homopolar motor? *Homopolar* essentially means "same polarity," so in a homopolar motor, the current is always moving in the same direction, from the positive terminal at the top of the battery to the negative terminal at the bottom, and the field it generates always has the same polarity.

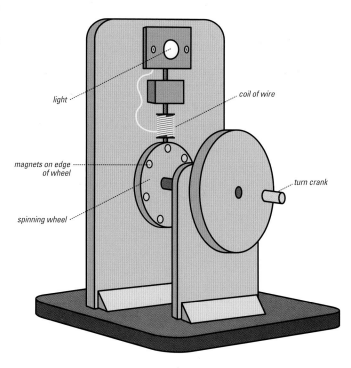

light
coil of wire
magnets on edge of wheel
turn crank
spinning wheel

A SIMPLE GENERATOR

MAKING MAGNETIC FIELDS VISIBLE

You can get a great look at the invisible magnetic field that surrounds a magnet by sprinkling iron filings on a piece of paper and putting the magnet underneath the paper.

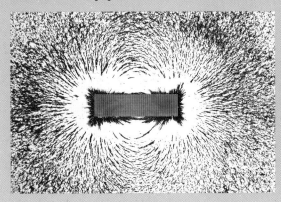

These iron filings create a pattern that makes visible the lines of force around the magnet and how they converge on two points, called the north and south poles. If you move two magnets close to each other, the forces are particularly strong when the poles are aligned: a north pole and a south pole are strongly attracted to each other, whereas two north poles or two south poles repel each other. With the neodymium magnets, these forces are so strong, they can painfully pinch your fingers if they get caught in between.

A motor in an electric fan or a toy car is not homopolar. These more complex and more powerful motors use multiple coils of wire to provide more force. An electrical switch called a *commutator* rapidly changes the direction of the current within those coils. That means that the magnetic field around each coil is rapidly changing polarity. The motor harnesses the power of these changing magnetic fields to make the rotor spin.

For more ideas, videos, and variations, visit www.eepybird.com/experiments/magnetmotor.

The Air Vortex Cannon is a giant gun that can shoot a blast of air all the way across a gym. Spook your friends with this invisible surprise, which can mess up their hair or blow out a candle. And then, you can make the invisible visible by sending smoke rings swirling across the room.

The air blaster most commonly in stores today, the Airzooka, was developed by Navy Lieutenant Brian Jordan. Jordan had been working on the idea ever since he was young and built his first prototype for it when he was still in high school. He kept refining the design over the years and finally put it on the market in 2003 where it was an almost immediate hit.

We'll show you how to make your own Airzooka-sized "pistol" from a plastic bucket, how to lock down your pistol for consistent aim, and how to make a giant "cannon" out of a trash can.

HOW DOES IT WORK?

Perhaps the most surprising aspect of the Air Vortex Cannon is that while the puff of air it shoots is powerful enough to make your hair fly up or knock over a stack of paper cups across quite a long distance, the puff of air also moves very slowly across the room. From the right distance it can take as much as two or three seconds before it hits its target. How can this cannon create such a powerful yet slow burst of air that travels so far?

When a bullet shoots out of the barrel of a gun or an arrow flies from a bow, they slice through the air without much adverse effect from the surrounding atmosphere. Solids are such dense, cohesive forms that they pass through air and other gases relatively freely. They are slowed somewhat by the friction caused by traveling through the air, but they keep their shapes as they fly along.

True to its name, the Air Vortex Cannon shoots air, which is no denser or more cohesive than the air that it's trying to move through. Unlike the bullet or arrow, you'd expect the puff of air from the cannon to dissipate, to melt away into the surrounding air. After all, no matter how hard you wave your hand around, you won't create air movement that will be noticeable across the room. And yet, tap your hand on the back of the Air Vortex Cannon, and your friends 20 or 30 feet away can feel the effects.

The key is the shape that puff of air makes when it leaves the cannon. It's a swirling donut called a toroidal vortex, that, as we'll see, has some very peculiar properties.

MATERIALS

The Pistol

- one 5-gallon plastic bucket (available at hardware stores)
- 1 heavy-duty large rubber band, approximately 5" diameter, 15" circumference
- 1 eye hook
- one 18" x 18" piece of plastic sheet (4-6 mm thickness; a plastic shower curtain liner works well)
- duct tape
- 1 piece of scrap wood (a short length of 2" x 4" will do)
- 1 golf ball
- one 9" length of 1" x 3" strapping
- one 6" length of 1" x 3" strapping
- one 2" drywall screw
- two 1¼" drywall screws

The Base and Target

- one 24" piece 1" x 3" strapping
- two 4" pieces of 1" x 3" strapping
- six 1¼" drywall screws
- 1 scrap of 2" x 4" approximately 6" long
- 6 paper or Styrofoam cups

TOOLS

- drill with ½" bit and ⅛" bit
- jig saw or cutting shears
- scissors
- channel lock pliers
- chop saw or handsaw
- screw gun or screwdriver
- 2 C-clamps (for base)
- 2 tables of the same height (for base)

HOW TO BUILD IT

STEP ONE: Mark the center point of the bottom of the 5-gallon bucket, and draw a 5" circle around it. An old CD or DVD is a good template for the circle.

STEP TWO: Using the drill with the ½" bit, drill a hole in the bottom of the bucket inside the 5" circle, but near the edge of the circle. This hole will allow you to get your jig saw blade in through the plastic to start cutting the circle.

STEP THREE: Using the jig saw, insert the blade through the hole you just drilled and carefully cut out the circle you traced on the bottom of the bucket. If your bucket is not too thick, you may be able to use cutting shears instead of a jig saw, but most of the buckets we have used have been too thick for shears to cut easily.

STEP FOUR: Using the drill with the ⅛" bit, make a hole in the side of the bucket approximately 8" from the top rim. Make a second hole on the opposite side of the bucket at the same height.

STEP FIVE: Using scissors, cut the rubber band once to make a single piece of elastic approximately 15" long. Reach inside the bucket and push one end of the elastic out through one of the ⅛" holes you drilled in the side of the bucket. Tie a large knot in the end of the elastic on the outside of the bucket so that the elastic cannot pull back through the hole.

Thread the other end of the elastic through the eye hook and then through the second ⅛" hole on the opposite side of the bucket. Again, tie a large knot in the end of the elastic on the outside of the bucket so that it cannot pull back into the bucket. The elastic should now stretch across the inside of the bucket with the eye hook hanging on the elastic in the middle of the bucket.

STEP SIX: Using scissors, cut the 18" x 18" piece of plastic sheet.

NOTE: We don't recommend using regular, 1.2-2-mm thick trash bags for the plastic sheet; these typically aren't strong enough to last.

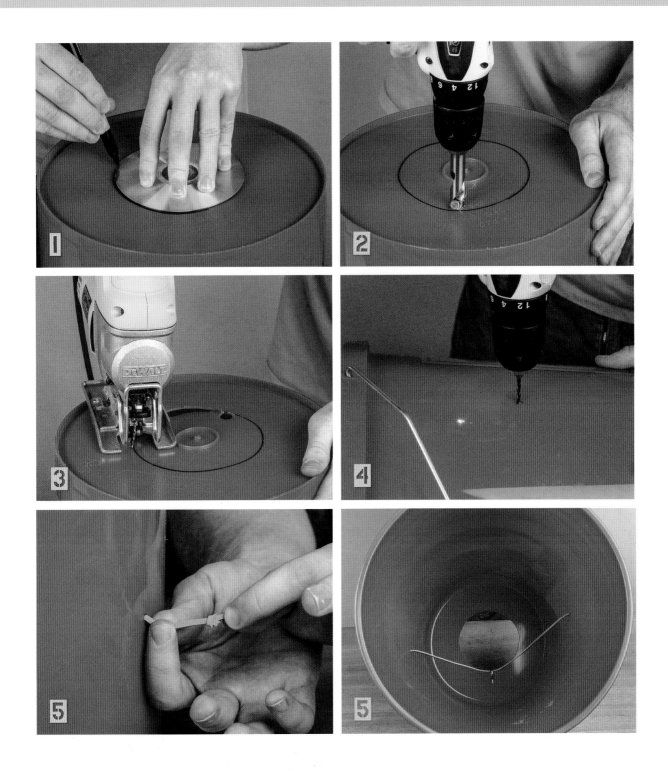

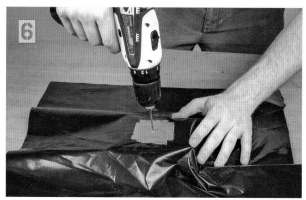

In the center of the plastic sheet, layer two 2–3" pieces of duct tape on top of each other to reinforce the plastic in that spot. Turn the plastic over, and do the same on the other side, reinforcing the center with two more layers of duct tape.

Place a piece of scrap wood underneath the plastic and drill a ⅛" hole through the center of the plastic and the four layers of duct tape.

STEP SEVEN: Drill a ⅛" hole approximately ½" deep in the side of the golf ball. To do this, hold the golf ball with the channel lock pliers or in a vise. Do NOT hold the golf ball with your fingers. The drill bit could all too easily slip off and hurt your hand, so use those pliers.

STEP EIGHT: Reach into the bucket from the top and pull up the eye hook on the elastic. Push the screw of the eye hook through the hole in the center of the duct tape-reinforced plastic, and screw the eye hook into the hole in the golf ball.

STEP NINE: Pull the edges of the plastic sheet up over the edges of the open top of the bucket, but don't pull it too tight across the top. Allow the elastic to keep the plastic down several inches inside the bucket, as shown in the photos. Using scissors, trim off some of the excess plastic. Use duct tape to attach the plastic to the bucket, running strips of tape all the way around the top sides of the bucket to secure the plastic in place.

STEP TEN: Make and attach the handle. Using the saw, cut the 9" piece of 1" x 3" strapping and the 6" piece of 1" x 3" strapping. It's simplest to make all right-angle cuts, but if you cut one end of the 6" piece at an angle of 20 to 30 degrees as shown in the photo, that will give your handle a comfortable angle like a pistol grip.

With the 2" drywall screw, screw through the center of the 9" piece and into the angled end of the 6" piece (if that's how you cut it) to make your pistol grip. Be careful with this step! Make sure you don't send the screw into your hand. If you angled the end of the short piece, make sure you start the screw into the 9" piece at the correct angle so that it will go straight into the short piece (as shown in the photo), not out the side where you're holding the wood.

Finally, place the 9" piece of strapping along the side of the bucket so that the handle angles back toward the plastic sheet and the golf ball. Attach the handle to the bucket with two 1¼" screws.

Done!

Test your Air Vortex Pistol by pulling the golf ball back towards the mouth of the bucket and letting it go. You should feel a burst of air coming from the circle you cut in the bottom of the bucket. Your Air Vortex Pistol is now ready to send air swirling all the way across the room!

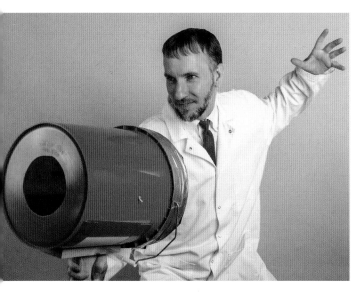

Building the Base and Target

The problem with the Airzooka has always been that aiming it, except in a very rough way, is nearly impossible. It's hard to know where that blast of air will hit. It's difficult to get a consistent shot because the vortex of air is invisible, and because there are so many variables with how to hold it, how to aim it, and how far back to pull the golf ball. Here's how to make *your* Air Vortex Pistol accurate: lock it down.

HOW TO BUILD IT

STEP ONE: Cut the 24" piece of 1" x 3" strapping and the two 4" pieces of strapping.

STEP TWO: Place the 24" piece on the table in front of you and put your Air Vortex Pistol (with the handle up and out of the way) across the center of the board, with the golf ball toward you. Place the two 4" long pieces of strapping snugly under it, lying flat on each side so that the bucket doesn't roll left or right.

STEP THREE: Mark the location of each small piece of strapping, remove the Vortex Air Pistol, and screw each small piece down to the 24" piece where you marked them, with two 1¼" screws on each.

STEP FOUR: Turn all of this upside down, so that the 24" strapping goes across the pistol, with the pistol in between the 2 small pieces. Using two 1¼" screws, screw the board onto the bucket to secure the pistol to the base.

STEP FIVE: Turn the pistol over so that the board is on the bottom and the handle is on the top again. Clamp the board to the edge of one table with the 2 C-clamps. This will lock down your firing angle, for consistent shooting. Surprisingly, clamping it down is important. Firing the gun jars it enough to throw off your aim. Slide the scrap of 2" x 4" under the front end of the Air Vortex Pistol as needed to precisely adjust the angle of the pistol.

STEP SIX: Place the other table about 10' away and stack the 6 cups into a pyramid (in ascending rows of three, two, and one on top).

STEP SEVEN: Adjust your tables and the angle of the Air Vortex Pistol so that it is aimed straight at the cups. Pull back the golf ball and fire! The air vortex should shoot out and knock down the pyramid. If your first shot isn't a direct hit, adjust left, right, up, or down as necessary. If sliding the 2" x 4" forward or back under the front of the pistol isn't enough to adjust the angle properly, loosen the clamps and add little strips of thin wood underneath the base (either in front or back) to adjust the angle even more; then tighten the clamps again. Once you've locked on your target, you should be able to hit it again and again. And don't stop with just paper cups. What else can you do? Can you blow out a candle from across the room?

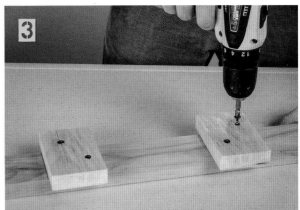

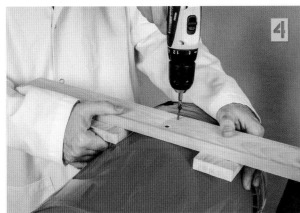

THE EXPERIMENT: THE GIANT AIR VORTEX CANNON

Now we'll make it bigger! For the Giant Air Vortex Cannon, the mechanism can be a little simpler: we won't need the elastic and golf ball.

MATERIALS

- 1 large plastic trash can (any size)
- 1 sheet of plastic (4-6 mm thickness; a plastic shower curtain liner works great)
- duct tape
- one 6' length of cloth or leather strap
- 1 small screw

TOOLS

- drill with ½" bit
- cutting shears or jig saw
- screw gun or screw driver
- scissors

HOW TO BUILD IT

STEP ONE: Mark a 10" diameter circle on the bottom of the trash can. You can trace around a dinner plate or paper plate to mark where to cut the hole—they're usually about the right size. Using the drill with the ½" bit, drill a hole in the bottom of the trash can inside the circle you marked to allow you to start cutting the circle out with the shears or jig saw. Unlike the smaller bucket, where we preferred using the jig saw, we've found that many trash cans can be cut with shears. Either with the shears or the jig saw, cut the 10" hole in the bottom of the trash can.

STEP TWO: Stretch the plastic sheet completely over the open top of the trash can (not the cut hole), leaving it a little bit loose, and trim any excess plastic around the edge with scissors. Secure the plastic sheet around the edge with strips of duct tape, as if the plastic were the head of a drum, but again, keep it a bit loose—don't stretch the plastic too tight.

STEP THREE: Attach the fabric strap, which will be a shoulder strap so that you'll be able to hold your Giant Air Vortex Cannon on your hip while you fire it. First, tie one end of the strap to a handle on the trash can. Then hold the cannon in place on your hip (you may need help) and confirm the correct strap length so that it will sit correctly once it's on your shoulder. Then, using 1 short screw, attach the strap to the front end of the cannon.

Now, give the center of the plastic sheet on the back a firm whack with the flat of your hand, and an air vortex will come shooting out the front.

As with the smaller version, it can be tricky to aim, but your Giant Air Vortex Cannon is ready to surprise people with invisible blasts of air—or add smoke and make those blasts visible!

It turns out that the air shot out of your Air Vortex Cannon doesn't travel straight ahead like you would expect. It twists back on itself, like an oddly shaped tornado, or what's called a *toroidal vortex*. The air moves quite quickly as it swirls around, but the vortex itself moves fairly slowly across the room.

A vortex is an area of a fluid (that is, a liquid or a gas, and in more advanced physics, a plasma) that spins around an axis. A tornado is a vortex of air. A hurricane is also a vortex, spinning around its eye. The water that swirls down the drain of your bathtube is a vortex.

So what is a toroidal vortex, and how does it form? *Toroidal* simply means "torus-like," and a *torus* is how mathematicians refer to a donut shape.

When you send a burst of air out of the end of the Air Vortex Cannon, the fast-moving air hits the surrounding air and starts to spread out. It first flattens into a widening disc of moving air, much the same way a cylinder of soft clay deforms if dropped on a cement floor.

After flattening, the air around the disc's edges starts to curl back on itself as the center of the disc pushes forward through the surrounding air.

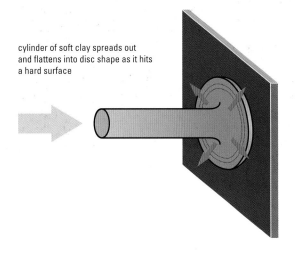

cylinder of soft clay spreads out and flattens into disc shape as it hits a hard surface

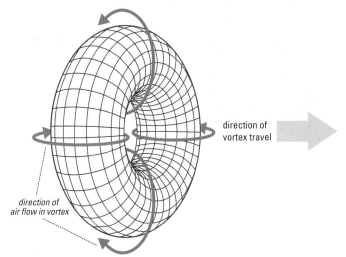

direction of vortex travel

direction of air flow in vortex

That circling air then gets pushed forward again by the air that's continuing to come out of the barrel of the cannon, and soon this process creates a spinning, donut-shaped air vortex. Like a tiny tornado, the air moves quickly *within* the torus shape, just like the wind in a tornado's funnel cloud moves at incredibly high speeds. Despite its high winds, however, the tornado itself moves relatively slowly across the landscape, and similarly, the spinning toroidal vortex from the air cannon only drifts forward slowly.

The cannon sends out just a quick burst of air. A continuous stream of blowing air would create a quite different effect. With a more sustained flow of air, a toroidal vortex would form on the leading edge, and the rest of the streaming air would follow in a roughly cylindrical shape. You can see this cylinder and donut combination in the familiar mushroom cloud after many explosions, including those caused by nuclear bombs or volcanic eruptions, where there is a large flow of gas, dust, and debris.

For more ideas, videos, and variations, visit www.eepybird.com/experiments/airvortex.

FIRING SMOKE RINGS

If you've always wanted a fog machine, this experiment makes a great excuse to get one. Small fog machines that generate a dramatic foggy mist from oil or glycol cost about thirty or forty dollars. You can find them online and from party and musician supply stores. They're perfect for your Air Vortex Cannon and great for parties and Halloween. If you've got one, fill your air cannon with fog and then start firing smoke rings.

If you don't have a fog machine, you can use incense logs, which are short cylinders of incense about ¼" in diameter and 2" long (don't use incense sticks; they don't generate enough smoke). Place 3 or 4 incense logs into the Air Vortex Pistol or Giant Air Vortex Cannon and light them. Cover the cut opening with a piece of paper and some tape, and leave it covered for 5 to 10 minutes. Once the inside is filled with smoke, uncover the opening and start firing.

One of our most popular videos transformed 250,000 sticky notes into giant waterfalls and wheels, using 3" x 3" zigzag pads of sticky notes made for pop-up dispensers. Here's how you can produce those same effects. While you can make waterfalls as big as you'd like, you can build smaller versions with only a dozen or fewer pads.

HOW DOES IT WORK?

Zigzag pads of sticky notes will expand and contract like a spring. When we first started experimenting with them, our first thought was that perhaps a pad could flow like a Slinky. As you'll see, yes, zigzag pads can actually walk down steps just like a Slinky—although it took a lot of time to figure out exactly how to get it to work smoothly.

Once we discovered how these pads could flow, our biggest goal was to combine lots of pads into giant waterfalls. But there was a problem. It turns out that different colored pads of sticky notes flow at different speeds: one color of green will flow faster than yellow which will flow faster than a different color green, and so on. We haven't been able to verify this with scientists at 3M yet, but we suspect different colored dyes saturate the paper more and give the paper slightly different stiffness, which then gives the pads different springiness. And that changes how quickly they flow.

To create waterfalls that would change color as they flowed, we had to figure out a way to synchronize pads flowing at different speeds. As you'll see, the answer was surprisingly simple, but as with many things that seem obvious once you think of them, it can take a long time to get there.

THE BASIC WATERFALL

Let's start with the Basic Waterfall, which uses just one 3" x 3" zigzag pad of sticky notes. The simplest trick is to get it to flow back and forth from one hand to the other.

When we first started experimenting with sticky notes, we thought that they would flow off the front as shown in the photo below on the left.

But if you try it, you'll see that the sticky notes catch on each other, and it doesn't work very well. It took us several days to think of rotating the pad 90 degrees and letting them fall off what's normally the side of the pad instead of the front or the back. When you do that, the notes flow perfectly as you can see in the photo on the right.

NO FLOW

FLOWS PERFECTLY

You can link as many sticky-note pads together as you'd like to make a Big Waterfall. Use one color or combine several, so that the colors change as they flow. Here's a waterfall you can build with 12 pads. We'll show you a waterfall that uses 3 different colors, but you can use whatever color combinations you like.

The key to getting Big Waterfalls like this to change color as they flow is using what we call synchronizing tape. This is simply clear Scotch tape that you use to connect the row of pads so that everywhere there's synchronizing tape, the pads are forced to flow together.

MATERIALS

- 12 zigzag pads of 3" x 3" sticky notes, made for pop-up dispensers
- Scotch tape
- glue stick

HOW TO BUILD IT:

STEP ONE: Set 4 pads of sticky notes next to each other, side by side, and oriented so that they all zigzag from side to side, like this:

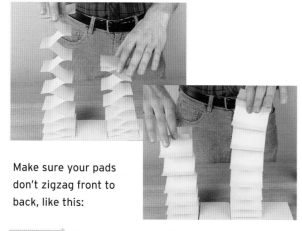

Make sure your pads don't zigzag front to back, like this:

STEP TWO: Put a piece of tape that extends across the top of all 4 pads. Use the glue stick and put down a layer of glue that covers the entire top of each of the 4 pads.

STEP THREE: Add another layer of 4 pads. First, peel off each pad's bottom paper (the nonsticky protective sheet with the company logo) so that the new pad will stick to the pad below it. Always make sure that every pad is oriented in the same way, so that it zigzags from side to side. Press each new pad firmly onto the glue you placed on the tops of the previous layer of pads in STEP TWO.

Repeat this step with another, final layer of 4 pads, so that each stack is 3 pads tall.

STEP FOUR: When all the pads are in place, put a couple more pieces of tape across the top of all 4 pads for strength.

STEP FIVE: Now carefully flip over your "brick" of sticky notes. Peel off the bottom papers (those sheets with the company logo). Then take the bottom sticky note off each pad, flip it over, and stick it back on so that there's nothing sticky showing. Once again, put a few pieces of tape across all 4 pads for extra strength.

Your brick is complete!

STEP SIX: It's time to get your waterfall flowing. Whether you have a 1-pad Basic Waterfall or a 12-pad Big Waterfall, you can get the notes to flow "over the cliff" using a box, a stack of books, a coffee table, a step, or any low, flat, raised surface. Set your waterfall on top of one of these raised surfaces. Grab the top sticky notes on each end of the waterfall, hold them tight, and guide the top of the waterfall down to the surface below. The rest of the waterfall will follow!

TIPS: Any size waterfall will flow down from a height of about 6" or 8" without any problem. To flow off something higher—and you can successfully go from as high as 2' with the Big Waterfall—all sizes of waterfalls need some guidance, or they will tip to one side or the other.

We've found that two straight-sided coffee mugs, drinking glasses, or small boxes do the trick perfectly. Just place one mug (or something similar) on either side of the waterfall, which keeps the pads aligned and straight as they flow down over the edge.

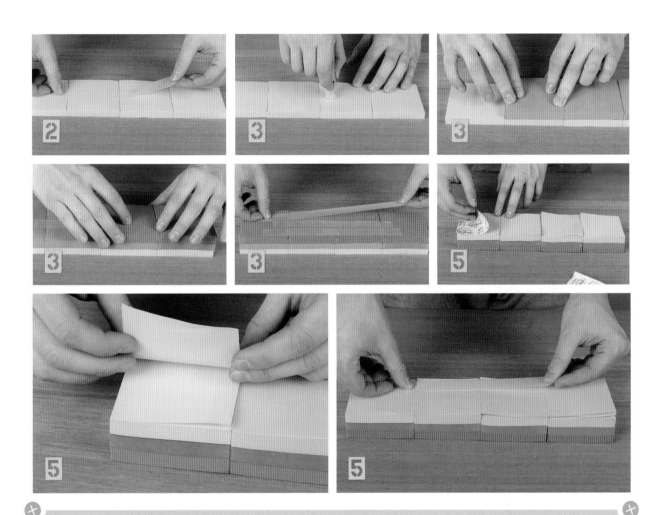

DISASTER AVOIDANCE TIP!

When moving your Big Waterfall, it's crucial that you carry the block by both ends. Grip each end firmly to lift it up. Like this:

If you carry the block by the middle, which is such an easy mistake to make, your waterfall may sadly end up looking like this:

Getting a Sticky-Note Slinky to walk down steps like a real metal Slinky is surprisingly tricky. You have to get the size of the steps and pad just right in relation to each other. Standard residential steps are too deep, so you must create shallower steps just for your Sticky-Note Slinky. In particular, you may need to adjust the step depth (called the "run"), so the Slinky lands on the same spot on each step, not too far out and not too far in.

In addition, the sticky notes can be temperamental. For instance, humidity is a problem. If it's humid, the sticky notes become heavier and less springy. Use a fresh pad, right out of the package, for more predictable flipping.

The method described here seems to give the most consistent results.

MATERIALS

- 1 zigzag pad of 3" x 3" sticky notes, made for pop-up dispensers
- duct tape
- 3 pieces of 11" x 6" cardboard
- Scotch tape

TOOLS

- utility knife
- 4-drawer filing cabinet (or 4 cardboard boxes)

HOW TO BUILD IT

STEP ONE: Split a single pad in half (each half will be a stack of about 50 individual sticky notes).

STEP TWO: Take the bottom sticky note off of each half pad, flip it over, and reattach it so that there's no exposed adhesive on the bottom.

STEP THREE: Place the two half pads next to each other with the zigzags going side to side, as shown in the picture. Put a 5" piece of duct tape across the top of the pads, linking them together. Add a second strip of duct tape on top of the first. You're using the duct tape both to hold the two pads together and to provide a little bit of extra weight on the ends of the pads.

STEP FOUR: Flip your double-wide pad over and put another 5" piece of duct tape across the bottom. Again, add a second strip of duct tape on top of the first for extra weight. This is your Slinky—short and wide—which we've found makes it much more consistent than tall and skinny.

Building the Steps

One of the easiest ways to build adjustable steps is to use a filing cabinet. The drawers are the right size, and a little cardboard turns them into usable steps. If you don't have a filing cabinet, you can use a stack of 4 cardboard boxes that are each about 12" tall.

STEP ONE: Using the utility knife, cut 3 pieces of cardboard to 11" x 6". Pull out the top drawer of the filing cabinet about 5". Place one of the pieces of cardboard with the 11" side running across the top edge of the drawer and tape the other 11" edge to the cabinet itself with Scotch tape. The cardboard should rest horizontally on the drawer, with about a 1" lip sticking out just over the front of the drawer as seen in the picture.

STEP TWO: Pull out each of the next 2 drawers about 5" farther out than the one above. Place one of the 11" x 6" pieces of cardboard across the top edge of each drawer as before, with a 1" lip extending over the edge and tape the other 11" edge to the bottom of the drawer

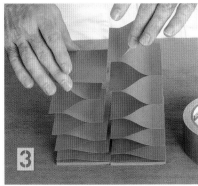

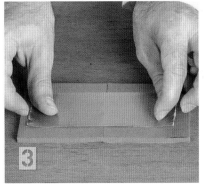

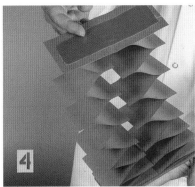

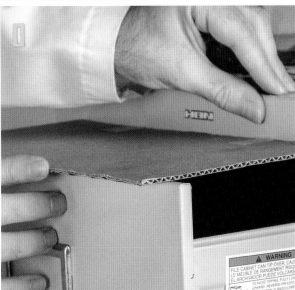

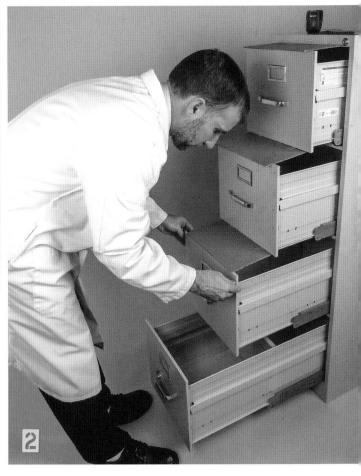

above with Scotch tape so that together, the pieces of cardboard make a set of 3 steps. The 1" lips help the sticky notes jump out past the filing cabinet's handles. If your file cabinet has recessed drawer handles, you don't need the 1" lip.

STEP THREE: Empty the bottom drawer and pull it out all the way. Don't put any cardboard across the top of this drawer—when the sticky notes jump down to the bottom, they'll fall right into the drawer.

STEP FOUR: Set your double-wide, half-height Sticky-Note Slinky on the top step and start it walking down the steps, as pictured on page 118.

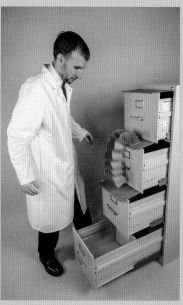

IDEAS TO TAKE IT FURTHER

Check out our videos at EepyBird.com for just how crazy this can get. How big can you make a Sticky-Note Waterfall? We haven't found the limits yet.

A fun extension of the Basic Waterfall is to make what we call the Herd. With a bunch of coffee mugs, glasses, or small boxes, you can set up a bunch of single-pad waterfalls and get them flowing at the same time.

What crazy color combinations and patterns can you make in a Big Waterfall? Try splitting pads into quarters to make shorter minipads (about 25 sticky notes per minipad), so that you can change colors more often and create more elaborate patterns. We've found that, given how fast the waterfalls flow, changing colors after every 25 sticky notes is about as fast as you can change colors and still perceive the patterns as they flow by.

VARIATION: THE PINWHEEL

The pinwheel is one of the easiest and most fun variations to make.

MATERIALS

- 1 zigzag pad of 3" x 3" sticky notes, made for pop-up dispensers
- glue stick

HOW TO BUILD IT

STEP ONE: Count off 26 sticky notes (that's 13 pairs— and it's easy to count them in pairs) and separate them from the rest of the pad. That's just over a quarter of the pad.

STEP TWO: Take one end and glue it to the other end using the glue stick. This makes what we call a starfish. It's almost a wheel, but it's not strong, so it won't roll.

STEP THREE: The trick is to rotate the sticky notes 90 degrees, flipping the starfish open to become a wheel, as shown above.

Now the Pinwheel is strong enough to roll.

See what else you can build from these zigzag pads. We've made sculptures and puppets, starfish, and giant wheels. Keep exploring to see what you can find!

THE SCIENCE

These experiments are dependent on the unique stickiness of sticky notes. So what's so special about how they stick to each other? Post-it notes are a classic story of turning an invention that looked like a failure into a huge success by finding the right application. 3M took a wimpy adhesive and turned it into an indispensable icon.

In the late 1960s, Spencer Silver, a scientist at 3M, which also makes Scotch tape and adhesives for airplanes, was trying to develop new polymers that would make better, stronger adhesives. He tried mixing an "incorrect" combination of various small molecules, but what he got was so weak it seemed useless.

However, the new adhesive had some interesting properties. It was only strong enough to hold two sheets of paper together, and it was so weak that the sheets could be separated without any harm. Plus, the stickiness remained, allowing the sheets to be stuck to other things or to each other multiple times, but without leaving any sticky residue.

The usefulness of these properties eluded 3M, and so the question remained: what do you do with wimpy glue?

Then, several years later, another 3M scientist, Arthur Fry, noticed how his bookmarks kept falling out of his hymnbook. He remembered hearing about Silver's weak adhesive, and he wondered if it could be used to create a bookmark that would stay in place until you wanted to move it.

It took another five years to perfect the idea and debut what would become the hugely popular Post-it notes. Today, sticky notes are as essential to the workplace as staples, paper clips, pencils, and even paper itself. All because someone recognized that this failed glue was really a unique adhesive in search of a purpose.

For more ideas, videos, and variations, visit www.eepybird.com/experiments/stickynotes.

This project contains two super-cool paper airplane launchers: the Sharpshooter and the Squadron Launcher. First, you'll learn how to build the Sharpshooter paper airplane. This is a variation on the basic dart, but with a few modifications so that it works well with a mechanical launcher. Then you'll build the Sharpshooter Launcher, which can fire an ordinary paper airplane over 60 feet with remarkable accuracy. Finally, you'll build the Squadron Launcher, which fires 10 planes at a time.

HOW DOES IT WORK?

Hold a piece of printer paper flat in your hand, then pull your hand away and watch the paper float slowly to the floor.

Now crumple up that same piece of paper and drop it from the same height. It falls more quickly. Why the difference?

As it falls, the flat piece of paper hits a lot more molecules of air than the crumpled paper does. Hitting air molecules is what causes *air resistance*, which slows any falling object. The wide paper encounters lots of resistance, so it dips and floats slowly, while the crumpled paper has much less surface area. It hits far fewer air molecules, encounters less air resistance, and thus falls faster and straight down.

When you make the classic dart paper airplane, you're folding the paper in order to manipulate the air resistance in a particular way so the plane can glide for as long as possible. You want it to float down slowly like a sheet, but with directional control, and you want it to move through the air with the speed of the crumpled ball, but without simply falling to the ground.

In the classic dart design, notice how much more surface area there is when looking at the plane from the bottom than from the front. This means the plane encounters more air resistance as it falls than it does as it flies forward, so it moves easily going forward through the air, and it resists falling down.

To get started, a paper airplane needs a push. With this experiment, we use rubber bands to supply the first burst of forward momentum.

MATERIALS

- 8½" x 11" printer paper (as many sheets as you want airplanes)
- paper clips (as many as you have airplanes)
- Scotch tape

multiply materials for the launcher below by 10 for the Squadron Launcher

- one 2' length of 1" x 3" wood strapping
- two 3½" pieces of 1" x 3" wood strapping
- one 6" long scrap of 2" x 4" lumber or similar block of wood
- four 1¼" drywall screws
- 2 eye hooks (size 10 works well)
- one 4" length of ½" diameter wooden dowel
- wood glue
- 2 medium to large rubber bands
- duct tape
- one 2" cotter pin

- two 6' lengths of 1" x 3" strapping
- two 9" lengths of 1" x 3" strapping
- forty 1¼" drywall screws
- four 2" drywall screws
- 10' length of cord or string

TOOLS

- handsaw or chop saw
- screw gun or screwdriver
- drill, with ½" and ⅛" bits
- square
- scissors

HOW TO BUILD IT

The Sharpshooter Paper Airplane

STEP ONE: Fold an 8½" x 11" sheet of paper in half lengthwise.

STEP TWO: With the lengthwise fold at the bottom (closest to you), fold down the upper right corner of the top layer, so that the bottom edge of the folded corner is flush with the lengthwise fold itself.

STEP THREE: Turn the paper over and mirror this fold on the other side.

STEP FOUR: On both sides of the paper, make a second diagonal fold in the same place. Using the diagonal folds you made in STEPS TWO and THREE, fold each side again so that the bottom edge is flush with the original lengthwise fold.

STEP FIVE: In the same place on both sides, make a third diagonal fold in exactly the same way, from the tip to back, aligning the bottom edge so it's flush with the original lengthwise fold. You have a plane!

STEP SIX: Bend open one side of a paper clip so that the bend at the top of the clip makes a 45-degree angle.

STEP SEVEN: Open the center of the plane and push the end of the paper clip through the base of the center fold of the plane about 3 inches from the tip of the plane so that the opened end sticks out at an approximately 45-degree angle. Tape the paper clip to the inside of the plane to hold it in place.

STEP EIGHT: Push the two sides of the plane closed and use a small piece of tape across the seam to hold it closed.

Repeat this process to make as many paper airplanes as you want. You will need 10 planes for the Squadron Launcher.

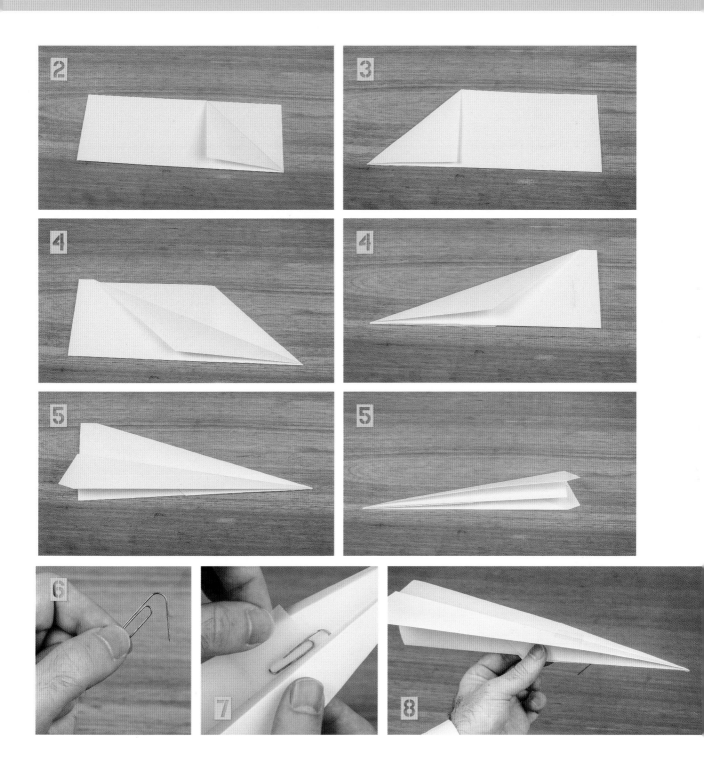

THE NAKAMURA LOCK

While today there are hundreds of paper airplane designs, we've found that the classic dart, the first paper airplane most of us learned to fold in grade school, is the best for these kinds of launchers. However, if you're looking for an awesome plane to quickly fold and fly, try the Nakamura Lock, a beautiful improvement on the classic dart. This won't work as well for the launcher, since the hook is more difficult to insert with the lock design, but it's a fun plane to know how to construct!

Eiji Nakamura introduced some basic origami techniques to the classic dart and came up with what may have been the most significant improvement in paper airplane design of the twentieth century. Nakamura's introduction of the simple lock fold in the center of the plane holds it together snugly, while at the same time this moves the center of gravity to a more optimal location for flight. Here's how to make the Nakamura Lock:

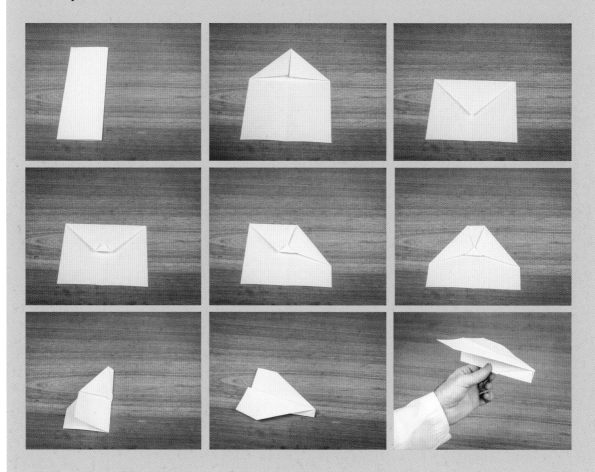

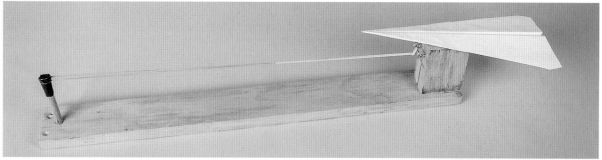

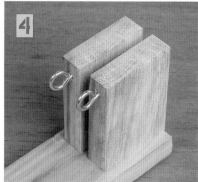

The Sharpshooter Launcher

The Sharpshooter Launcher is a simple-to-make, rubber band–powered paper airplane launcher. A chain of rubber bands is stretched across the launcher and held in place with a pin. When you're ready to fire, hook the plane's paper clip to the rubber band and pull out the pin; the rubber band releases and the plane shoots out of the launcher.

STEP ONE: Using the chop saw, cut a 24" length of 1" x 3" wood strapping. This will be your base.

STEP TWO: Cut two 3½" pieces of 1" x 3" wood strapping. These will be your plane supports.

STEP THREE: At one end of the 24" base, attach the 2 plane supports. The supports should be standing up on end, side by side, about half an inch apart from each other, with the gap aligned with the center of the base. When they are correctly positioned, attach the two plane supports from beneath with a screw gun and two 1¼" drywall screws in each piece of strapping.

STEP FOUR: On the front edge of each plane support (the edge facing the opposite end of the 24" base), just below the upper corner, screw in an eye hook by hand. Twist the eye hooks until the holes align vertically with each other, as shown.

STEP FIVE: Using the drill with the ½" drill bit, drill a hole in the opposite end of the 24" base, approximately 1–2" from the end. You may want to place a block of scrap wood underneath the point where you're drilling to protect the table underneath, as shown in the photo. Don't drill all the way through the base; stop about three-quarters in (it still may be a good idea to have that block of wood underneath in case you go too far). Then glue the ½" dowel into the hole with wood glue so that the dowel stands straight up in the hole.

STEP SIX: To complete your launcher, tie or knot 2 rubber bands together (use more if 2 aren't long enough to stretch from one end of the base to the other). See page 128 for a great way to link the rubber

5

5

HOW TO TIE
2 RUBBER BANDS TOGETHER

Start by stretching one rubber band (we'll call this the "top" rubber band) over the tips of your forefinger and thumb of one hand, and then hang the second rubber band (we'll call this the "bottom" rubber band) on the tip of your forefinger, as shown in the photo. With your other hand, reach through the bottom rubber band and grab the top rubber band, starting to pull it back through the loop. Now pinch the bottom rubber band between your forefinger and thumb, so

that as you move your hands apart, it makes the interlocked knot shown in the photo. Pull it

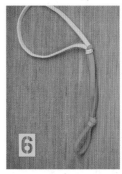

6

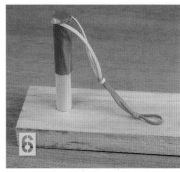

6

7

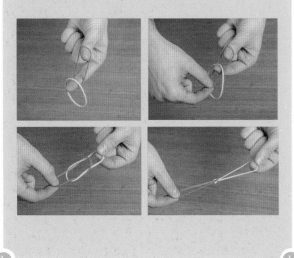

7   **7**

bands together. Hook one end of this chain of rubber bands over the top of the dowel, and use some duct tape to attach them to the top of the dowel. Then, at the other end of the rubber band chain, tie a knot to create a tied-off loop about 1" across. You'll eventually attach the airplane to this loop.

STEP SEVEN: To prepare the Sharpshooter for firing, pull the rubber band back and hold it in place with the cotter pin. To do this, insert your index and middle fingers into the large loop of the rubber band furthest from the dowel (but not the small tied-off loop from STEP SIX). Pull the rubber band back from the dowel so that it passes between the plane supports. Align the open rubber band with the 2 eye hooks and run the cotter pin through both eye hooks and the large loop of the

rubber band. Gently release the rubber band so it snugs against the cotter pin. Make sure the small tied-off loop is on top and sticking up, just above the eye hooks. If it's underneath, slide or adjust the rubber band until the small loop is positioned on top, ready for the airplane's paper clip hook.

When set, it should look like the photo below.

STEP EIGHT: Set the launcher on a table or on the floor. Place the scrap of 2" x 4" underneath the front of the base to tip it up. You can slide this block of wood back and forth or add more blocks underneath to adjust the angle of the launcher to whatever angle you want.

Make certain you have a clear space in front of you for at least 60' or 70' (or as far as possible before you hit a

TARGET PRACTICE

This experiment becomes even more fun when you set up a target. For our first target, we aimed for an open window at the far end of our workshop, 60' from where we were firing. It took a few tries, but we sent several airplanes straight through the window. At first, aiming is hit-and-miss, but the Sharpshooter is a very consistent launcher. Keep adjusting its position and improving your aim, and you'll soon be hitting the target.

wall), as well as on either side of the flight path. Planes don't always fly straight, particularly once they've been refolded after suffering minor damage on previous flights.

STEP NINE: Load the plane. Pinch the center portion of the plane together and slide it between the plane supports so that the wings are spread on top of the supports. Then hook the paper clip onto the small tied-off loop of rubber band above the eyehooks.

NOTE: It doesn't work as well if you hook the airplane onto the larger loop where the cotter pin is holding the rubber band under tension.

STEP TEN: Aim your plane into the clear space and pull the cotter pin, releasing the rubber band. When everything goes right, the plane will fly straight and true. If your plane takes off with too much force, add more rubber bands to your rubber band chain as needed.

The Squadron Launcher

The Squadron Launcher is the Sharpshooter times ten. The cotter pins are strung together so they can all be pulled at once.

STEP ONE: Make 10 Sharpshooter Launchers, just like the one above.

STEP TWO: Using the saw, cut the two 6' lengths of 1" x 3" strapping. These form the base of the Squadron Launcher to which you attach the individual Sharpshooter Launchers, as shown below.

Using the 1¼" screws, screw the individual launchers into the 2 crosspieces of strapping (as shown). Leave a 4½" space between the individual launchers. Use 2 screws, positioned on a diagonal, to attach each Sharpshooter Launcher to each crosspiece. Placing screws on a diagonal will help reduce twisting of the individual launchers and of the frame as a whole. As you attach everything, try to keep the launchers lined up evenly and all parallel, pointing straight ahead.

STEP THREE: Using the saw, cut two 9" pieces of 1" x 3" strapping. These two supports lift the front of the Squadron Launcher, so it's angled up. On the very end of the third and eighth Sharpshooter Launchers, drill two pilot holes with the ⅛" drill bit (this helps ensure the wood doesn't split), then, from the top and using two 2" drywall screws, screw into the ends of each 6" piece of strapping.

STEP FOUR: Make the firing pin string. Using scissors, cut a 9' length of cord or string.

Tie 1 cotter pin to the very end of the string. Then tie the next 9 cotter pins onto the string at 8–9" apart from each other.

These can be tricky knots to tie. Here's one knot that works well: At the point at which you want to tie the cotter pin onto the string, loop the string and push the loop through the hole on the pin. Pull that little loop through and then up and over the pin, pushing the point of the pin through the loop of string. Once the pin is through the loop, pull the string back down snug, and it will have made a nice knot.

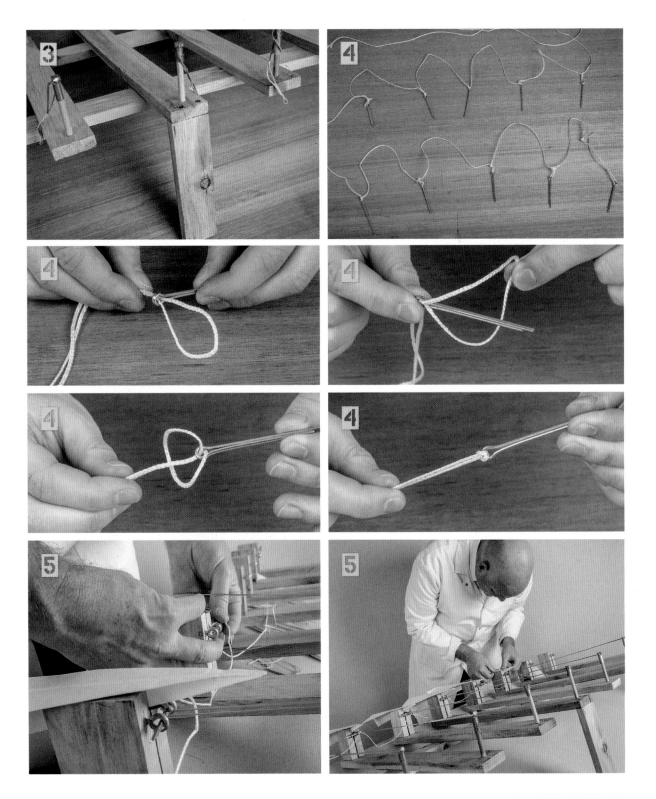

STEP FIVE: Lay the firing pin cord across the launcher, in front of the eye hooks, so that each set of eye hooks has a pin right next to it. Prepare the launcher by securing each rubber band with a cotter pin one at a time. Start at the end of the launcher from which you'll be pulling the cord and work your way toward the far end. As on the Sharpshooter Launcher, pull back the rubber band so it's between the plane supports and level with the eye hooks; insert the first cotter pin through both eye hooks and the rubber band loop. When doing this, it's very important to orient all of the cotter pins so that the heads (where the cord is attached) all face the side from which you'll be pulling. ("Reversed" cotter pins won't come through the eye hooks when pulled and the cord will jam.)

STEP SIX: Once all the individual launchers are ready for loading, double check the set up: are all of the cotter pins oriented in the right direction? Once they are, hook the planes in.

STEP SEVEN: Check that the area is clear in front of you, and pull the firing cord. Your entire squadron of 10 planes should fire at once.

THE SCIENCE

ow do planes, even paper airplanes, fly? Four primary forces must be correctly balanced: gravity, drag, thrust, and lift. The first two forces work against the plane and try to bring it down, while the second two forces help the plane overcome gravity and drag and remain in the air.

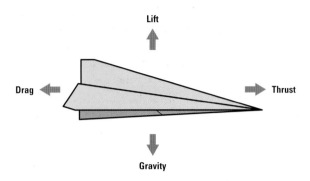

GRAVITY On Earth, gravity is always pulling everything toward the ground. Anything that flies has to overcome that force with an equal yet opposite force: something that pushes you up. This opposite force is called lift. If lift is greater than gravity, what flies will move up; if lift is less than gravity, what flies will descend. If an object generates virtually no lift at all, gravity wins and the object falls like a rock (or like a crumpled ball of paper). But in the Earth's atmosphere, a piece of paper falls a little bit more slowly than a rock because of . . .

DRAG Drag is the air resistance caused by anything that moves through the air, hitting air molecules, and knocking them out of the way. This takes energy and slows the plane down (just like it slows down the swirling puff of air from the Giant Air Vortex Cannon on page 108). Drag is also a factor in automobile design. Cars are built in more aerodynamic shapes (that is, shapes that minimize air resistance) so that cars can move faster with less effort

(that is, using less gas). A hundred years ago, we didn't know much about aerodynamics and designed cars very differently.

THRUST Thrust is whatever force propels the plane (or anything else) forward. With a plane, thrust comes from the propellers or the jet engines. With a paper airplane, thrust is provided by your arm or by the rubber band of your Sharpshooter Launcher. Thrust is what helps the plane overcome the drag of air resistance.

LIFT Lift is the upward force that counteracts gravity. Different things can generate lift. In an airplane, lift is generated by precisely shaped wings. The forward movement of a wing's profile creates lower air pressure above the wings and higher pressure below the wings as the plane moves forward through the air. If the air pressure above the wings is lower than the air pressure below them, then the air itself will essentially push up against the plane's wings. One way a pilot changes a plane's altitude, to go up or down, is by changing the angle of the wings and varying the amount of air pressure above them.

In other words, for a plane to keep flying, the lift forces must be greater than the force of gravity pulling the plane down. For the plane to move forward (and to create sufficient airflow for the wings to generate lift), the thrust forces must be greater than the air resistance that creates drag.

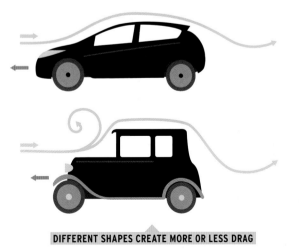

DIFFERENT SHAPES CREATE MORE OR LESS DRAG

Not everything that moves through the air "flies." A lot of things just fall slowly. Or not so slowly. When you throw a stick or a rock, the only thing keeping it in the air is the thrust of your throw. With nothing to add to this thrust, air resistance slows its forward momentum and gravity pulls it down faster and faster. A thrown stick of course moves forward *while* it falls, and if you throw it upward, it will ascend briefly as well, but it can't stay up in the air because the average stick just doesn't have the right shape to generate much lift. It has nothing to help it fight against gravity and drag except the power of your throwing arm.

That said, the dart-style paper airplanes used with the Sharpshooter Launcher don't generate a lot of lift either. They're designed primarily to minimize drag, or the air resistance encountered as the plane moves forward, and to maximize the air resistance encountered under the wings as the plane starts inevitably moving downward. This maximizes the effects of the thrust of your throw and extends the airplane's "falling." Yet, with the right shape, the wide wings can generate some lift. Other paper airplanes are designed to maximize lift and don't need to be thrown with much force at all; these are the ones that loop, turn, and dip on their own. Even the classic dart with its modest design, when given a good hard throw, can fly a remarkable distance in a straight line.

 For more ideas, videos, and variations, visit www.eepybird.com/experiments/squadron.

LEVEL III

THE BIG STUFF

DON'T FORGET!

Be safe! Going into Level III, this is a good time to remind yourself of A Word About Tools and Safety on page 68. When you need help, whether to use a power tool safely or just to hold something when you wish you had four hands, find someone. Keep all your fingers and toes in good working order. And wear those safety goggles!

15 THE LEAF-BLOWER HOVERCRAFT AND ITS LITTLE COUSINS

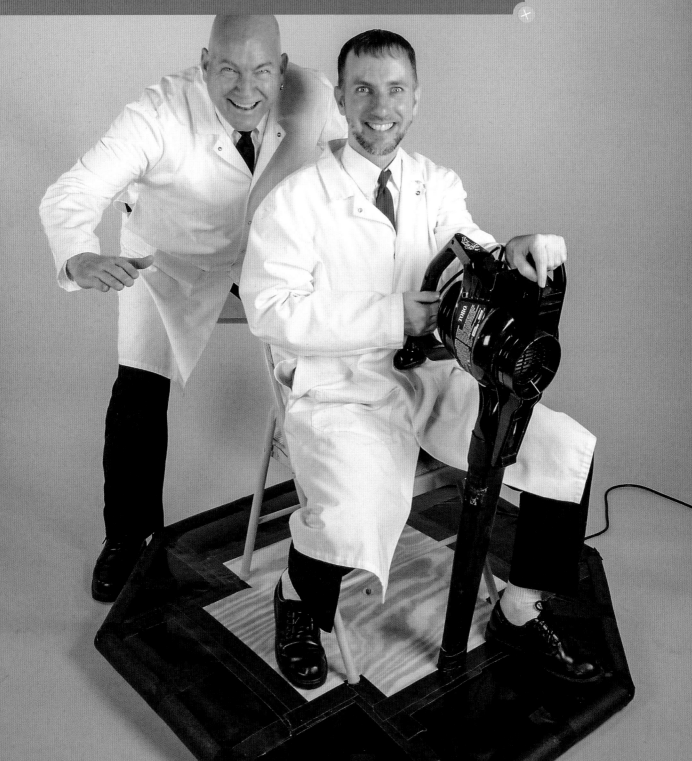

Have you ever dreamed of floating around on a cushion of air? Sound impossible? It's not! Using just a leaf blower, some plywood, and a sheet of plastic, you can!

We'll show you three different hovercrafts you can build, from small to human-sized. You get started with a tabletop version that uses a CD and a balloon. Then you use a hair dryer and paper plates. And then you move up to bona fide human transportation with your very own Leaf-Blower Hovercraft.

HOW DOES IT WORK?

When your hovercraft is turned off and just sitting on the floor, it's hard to move, obviously. The friction between the hovercraft and the ground prevents easy sliding. That's no fun at all.

Turn on the leaf blower, however, and the air forced out the bottom separates the hovercraft from the ground just enough so that they no longer touch. Now the hovercraft can slide almost friction free, lubricated by this cushion of air.

How does a hovercraft achieve liftoff? As we just saw in the Paper Airplane Squadron, airplanes are able to get off the ground because rapid forward motion pushes the air around the wings, making areas of lower air pressure above the wings and areas of higher air pressure under the wings. That difference in air pressure creates the force called *lift*, which, true to its name, is what lifts the airplane into the air. But a hovercraft doesn't have to be moving to rise off the ground. It creates lift just by pushing air down to make an area of high air pressure underneath.

You have to push a lot of air underneath a hovercraft to create enough lift. That takes a lot of power. It wasn't until the 1950s that an engineer named Sir Christopher Cockerell came up with a design that was efficient enough so that he could build large hovercrafts. Cockerell came up with the idea of a "momentum curtain" that uses a ring of air to create high pressure in the middle of the ring. You'll mimic this by attaching a skirt of plastic underneath your Leaf-Blower Hovercraft to help direct the airflow in the same way, so that the power of just a leaf blower is enough to lift a person off the ground.

THE EXPERIMENT: THE TABLETOP HOVERCRAFT

Let's start small, with a tabletop version that is quick and easy to build.

MATERIALS

- 1 pop-up bottle cap from a plastic drinking bottle or dish soap container (the kind you pull up to open and push down to close)
- 1 old CD or DVD disc
- glue
- 1 balloon

HOW TO BUILD IT

STEP ONE: Close the pop-up bottle cap so it's airtight. Then glue the base of the pop-up bottle cap directly over the hole in the center of the CD/DVD.

STEP TWO: Inflate the balloon, pinch it closed, and place the mouth of the balloon over the pop-up bottle cap. Twist the neck of the balloon a few times to make sure that no air escapes until you're ready to launch.

STEP THREE: Place your Tabletop Hovercraft on a smooth surface, squeeze the balloon neck, pull open the pop-up bottle cap to open it, and let go. As the balloon untwists, the escaping air will flow through the bottle cap and under the CD, creating a cushion of air. Your Tabletop Hovercraft will float freely, and you can push it around in any direction until the balloon runs out of air.

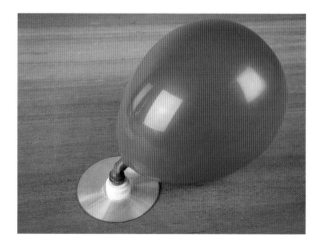

THE EXPERIMENT: THE HAIR-DRYER HOVERCRAFT

MATERIALS

- 3 paper plates (the sturdier the plate, the fewer you need)
- duct tape
- 1 hair dryer
- 1 pencil

TOOLS

- scissors

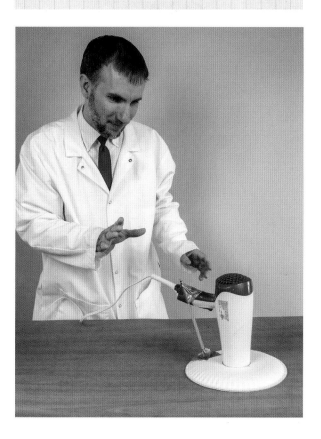

HOW TO BUILD IT

STEP ONE: Stack the paper plates. If you find you need added strength, duct tape them together around the edges.

STEP TWO: Trace the nozzle of the hair dryer in the center of the stack of paper plates. Then use the scissors to cut out the nozzle outline.

STEP THREE: Place the nozzle of the hair dryer in the hole and tape it securely in place. If the hole is tight enough, you may be able to just wedge the hair dryer in place.

STEP FOUR: Tape one end of the pencil to the handle of the hair dryer, then tape the other end of the pencil to the paper plate. This way, the pencil acts as a support for the hair dryer to keep it from falling over.

STEP FIVE: Place the hovercraft on the floor with the paper plates facing down, plug in the hair dryer, and turn it on full. The Hair-Dryer Hovercraft should float freely as far as the cord will allow (use a household extension cord to increase your range). If your hair dryer has a setting that just blows without heat, that's best. You don't need hot air, just lots of fan power.

If you have more than one hair dryer, you can build more than one Hair-Dryer Hovercraft. What can you come up with to do with two little hovercrafts?

THE EXPERIMENT: THE LEAF-BLOWER HOVERCRAFT

This hovercraft is made on a plywood base with a 6-mil plastic skirt on the bottom. The leaf blower inflates the skirt through a small hole in the plywood cut to fit the blower nozzle. The air then fills the skirt and escapes through six 2" holes on the bottom of the skirt. The air escaping those holes creates the cushion on which the hovercraft floats. This basic design goes back to at least 1987, when Barbara Saur of Kent, Washington, demonstrated a "human airpuck" in a course for physics teachers. Her design was written up in the November 1989 issue of *The Physics Teacher* and further developed by William Beaty of amasci.com. Now, even the Mythbusters have built their own hovercrafts.

MATERIALS

- one 4' x 4' piece of ½" plywood
- one 5' x 5' plastic sheet, 6 mil thick
- duct tape
- 16' of ¾" foam pipe insulation
- one 2" bolt, ¼" diameter
- 2 fender washers, approximately 2" diameter
- 1 plastic coffee can lid
- 1 nut, ¼" diameter
- 1 leaf blower (a corded electric leaf blower is best; cordless blowers are usually too weak, and gas-powered blowers aren't good for inside)
- extension cords (the length of these determines how far you'll travel)

A WORD ABOUT PLYWOOD

Standard plywood sheets are 4' x 8' and can be hard to manage and transport. However, many hardware stores and lumberyards stock precut 4' x 4' sheets (which is what this experiment calls for). If your hardware store doesn't, they'll probably be happy to cut a 4' x 8' sheet in half for a nominal charge. Have them do this for you, since it saves you the hassle.

Also, plywood comes in different grades, depending on the quality of the grain, the number of knotholes, and so on. Typically, each side is a different grade or quality of surface. For example, "B/C" grade plywood has one surface that is "B" grade and one surface that is slightly poorer "C" grade. For this project, the cheapest grades will do, but whatever you get, use the higher grade (and smoother) surface for the side facing the ground, since that will be less likely to catch on your plastic skirt and tear.

TOOLS

- permanent marker
- circular saw
- jig saw
- drill, with ⅜" or 5⁄16" drill bits
- staple gun with ⅜" or smaller staples
- scissors
- utility knife
- 2 wrenches

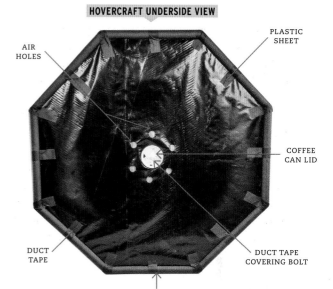

HOVERCRAFT TOP VIEW

HOLE FOR LEAF BLOWER

PLYWOOD

PLASTIC SHEET

BOLT

WASHER   NUT

FOAM PIPE INSULATION

DUCT TAPE

HOVERCRAFT UNDERSIDE VIEW

AIR HOLES

PLASTIC SHEET

COFFEE CAN LID

DUCT TAPE

DUCT TAPE COVERING BOLT

FOAM PIPE INSULATION

HOW TO BUILD IT

Cutting the Plywood

STEP ONE: Mark the center of your 4' x 4' piece of plywood on both sides with a permanent marker. The easiest way to do this is to draw diagonal lines from each corner to the opposite corner. If your piece is truly square, the center is the spot where the diagonals cross. For this project, you don't need to worry if the measure is off by a little, as exact dimensions aren't particularly important.

STEP TWO: Once you've marked the center on both sides, use the circular saw to trim off the corners so that you make an octagon roughly 20" (or 19⅞" if you want to be more precise) on each of the 8 sides. The easiest way is this: from each corner, measure 14¹⁄₁₆" in both directions, mark a diagonal line between these points, and then cut along this line. A circular saw is preferred for these cuts (as a jig saw is harder to hold on a straight line and a handsaw takes forever).

You want an octagon so that your hovercraft doesn't have sharp corners, which could hurt whatever it bumps into (including your toes!), and it *will* bump into things until you get the hang of flying it. If you're more

ambitious, and want a sleeker look, cut a circle instead of an octagon. To mark a circle, tack down a 2' piece of string in the board's center, tie a marker to the end, and use that as a compass. Then cut along the circle with a jig saw.

STEP THREE: Drill a hole in the center of the plywood that is slightly larger than the diameter of your bolt. We're using a ¼" diameter bolt, so drill a hole with a ⁵⁄₁₆" or ⅜" bit.

STEP FOUR: Cut out a hole in the plywood for the leaf-blower nozzle. To do this, mark a point 12" from what will become the front edge of the hovercraft and 24" from either side. Place the leaf-blower nozzle over this point and trace a line around it. Cut out this shape for the hole.

HOW TO CUT A HOLE IN THE MIDDLE OF A PIECE OF WOOD

To cut a hole in the middle of a piece of wood, start by drilling a hole (large enough for your saw blade) inside the section you want to remove. Then, insert your jig saw blade into the hole and start cutting from there. Saw to the edge of the nozzle hole you've marked. If necessary, cut out the hole in pieces rather than trying to cut the entire hole in one shot.

Attaching the Plastic Skirt and Bumper

STEP ONE: Spread the 5' x 5' piece of 6-mm plastic flat on the floor and lay the octagon of plywood flat in the middle of the plastic. Wrap the plastic over the edges of the plywood, and attach it to the top of the plywood with ⅜" staples. As you work your way around with the stapler, gently stretch the plastic so that it fits snugly over the bottom of the plywood.

Be generous with the staples. The plastic skirt should be well secured to the plywood. We recommend placing the staples about a half-inch apart. Also, note that you want to use small, ⅜" staples. If you use bigger staples, it's harder driving them all the way into the plywood.

STEP TWO: Once the plastic is securely stapled, trim the excess plastic with a pair of scissors, leaving about ½" margin past the staples.

STEP THREE: Duct tape the plastic to the plywood to completely seal all of the edges. This seal needs to be air tight, so do this carefully and don't skimp on the tape.

STEP FOUR Attach the bumper. With the utility knife, cut eight 20" lengths of foam pipe insulation and slide one of these onto each edge of the plywood. You can get pipe insulation that has adhesive already on it, so it will stick right to the skirt on the hovercraft, or you can use strips of duct tape to attach the insulation to the skirt. This will create a soft but sturdy bumper that will prevent your hovercraft from scratching or marring anything it bumps into. Did we mention the part where the hovercraft bumps into you?

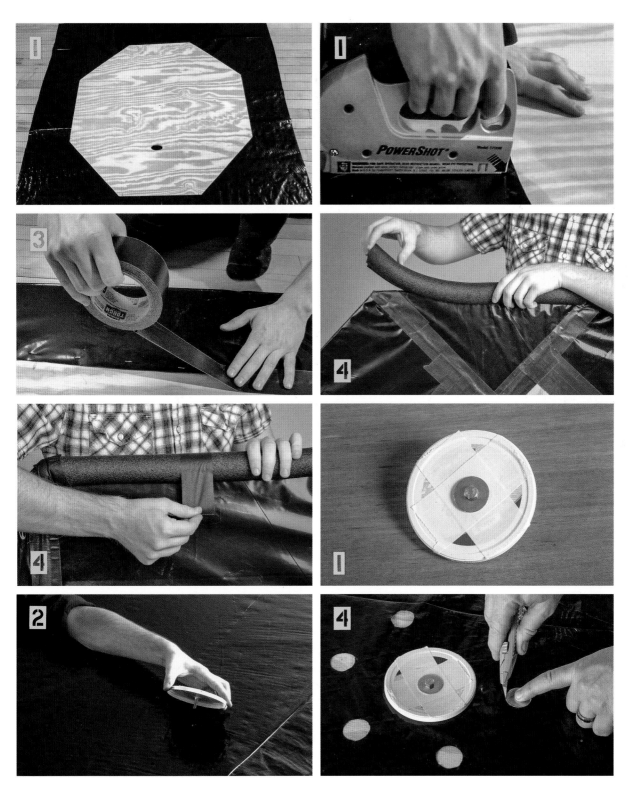

Creating the Holes for the Air Cushion

STEP ONE: Thread one of the fender washers onto the 2" bolt. Using the utility knife, cut a small "X" in the center of the coffee can lid, and thread that onto the bolt next.

STEP TWO: From the bottom of the hovercraft, push the bolt up through the center of the plastic skirt and through the hole in the center of the plywood.

STEP THREE: Thread the second fender washer onto the bolt from the top of the plywood and then screw on the nut. Tighten this with 2 wrenches from both sides to make sure it is securely fastened. The purpose of attaching the coffee can lid to the bottom is to hold the center of the plastic firmly to the hovercraft without tearing it. Put a couple pieces of duct tape across the head of the bolt onto the lid so that the head of the bolt won't scratch your floor if you slide the hovercraft.

STEP FOUR: With the utility knife, carefully cut six 2" diameter holes in the plastic sheet on the bottom. The holes should be approximately 6" apart and also about 6" from the coffee can lid, or the center of the hovercraft. If you have an extra fender washer, use that as a guide to cut around with the utility knife. These dimensions are approximate, but be careful not to place the holes too close together or the plastic between them may tear.

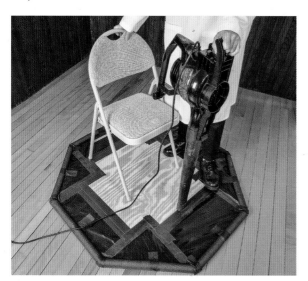

These holes will allow air to escape from the plastic bladder and create the cushion of air on which the hovercraft will float.

Flying Your Hovercraft!

STEP ONE: Place a light chair in the center of the hovercraft. A resin patio chair is ideal, but any light chair, such as a folding chair, will work. You can fly your hovercraft standing up, but it's harder to maintain your balance (standing raises your center of gravity), and it's easier to fall and hurt yourself. It's not necessary to attach the chair (or the leaf blower) to the plywood, and leaving them unattached makes the hovercraft easier to disassemble and store afterward.

STEP TWO: Slide the leaf blower nozzle into the nozzle hole you cut in the plywood. If the nozzle is loose, use duct tape to secure it in the hole and to the plywood; you want a tight fit. If the nozzle pops out by mistake, you'll just sink to the ground until you put the nozzle back in.

STEP THREE: Turn on the leaf blower! You'll rise up about 2" off the ground and float on a cushion of air. Have a friend give you a push, and you'll float nearly friction-free all over the room.

Don't skimp on the extension cords! The longer these are, the farther you can go. However, it can be helpful to have your friend keep the cord from getting tangled up as you float around.

STEP FOUR: Stopping the hovercraft: The braking system on this design is simple and effective. Just turn off the leaf blower. Or simply remove the leaf blower nozzle from the hole in the plywood. The bladder will instantly deflate and you'll come to a remarkably smooth stop.

NOTE: One common problem you may encounter riding the Leaf-Blower Hovercraft is centering your weight correctly. If you sit too far forward or too far back, the plywood will tilt and push one edge into the floor, preventing the hovercraft from moving. If this happens, keep shifting your weight in the chair till you're centered and the hovercraft is floating level.

All the way back to before the invention of the wheel, humans have tried to create ways to reduce *friction*. Before simple rolling mechanisms, if you wanted to move something from one place to another, you could drag it, push it, or lift it. That works fine for relatively light objects, but the heavier the object, the more trouble friction causes.

Friction is the resistance to movement that occurs when two objects rub against each other. (For more on this, see Yanking Tablecloths and Other Near Disasters, page 44.) Microscopically, friction occurs when each object's *asperities*—the small bumps and ridges that make a rough surface rough—bump into each other and keep the objects from sliding easily.

Of course, we can use this property to our advantage. If the asperities of one surface are hard (such as the grit on sandpaper) and another's are soft (like wood or wax), then the softer asperities can be knocked off and that surface will become worn down and smoother. That's what happens when you sand a rough piece of wood.

How Rubberized surfaces make things "grippy"

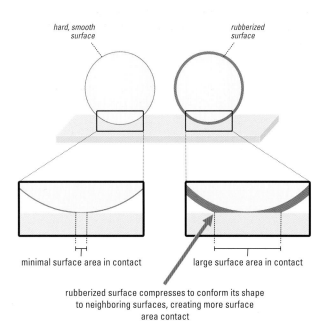

hard, smooth surface

rubberized surface

minimal surface area in contact

large surface area in contact

rubberized surface compresses to conform its shape to neighboring surfaces, creating more surface area contact

Friction with and without lubrication

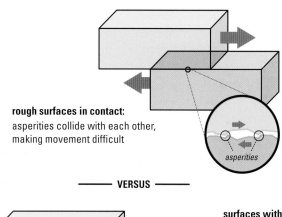

rough surfaces in contact: asperities collide with each other, making movement difficult

asperities

— VERSUS —

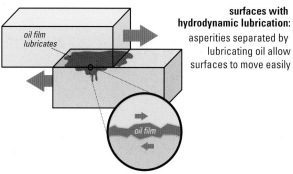

oil film lubricates

surfaces with hydrodynamic lubrication: asperities separated by lubricating oil allow surfaces to move easily

oil film

Often, you *don't* want surfaces to slide past each other. Instead, you want friction to give you a better "grip." If you've ever tried to make your way across a patch of ice on the sidewalk, you know how important a little friction is even for doing things as basic as walking.

A rubber surface is grippy precisely because it's soft enough to conform around the other surfaces pressed against it. This increases the surface areas in contact with each other, which creates more friction and a better grip.

When both objects are hard, and yet we want them to move more easily past each other, we often introduce *lubrication*, such as oil. The lubricant gets in between the asperities and separates them just enough so that they don't hit each other. This is how we make the pistons in our car engine slide more easily. And that's why, if we run out of oil in the engine, the parts seize up, unable to overcome the friction.

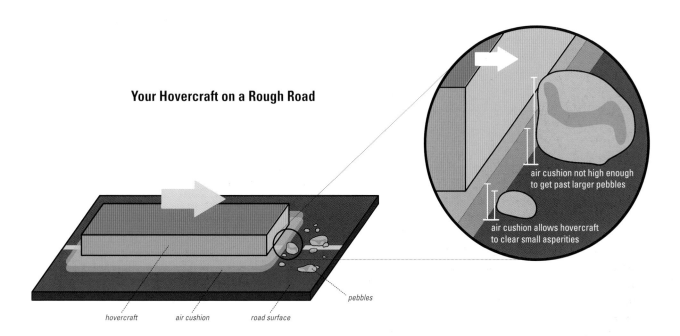

Your Hovercraft on a Rough Road

air cushion not high enough
to get past larger pebbles

air cushion allows hovercraft
to clear small asperities

pebbles

hovercraft   air cushion   road surface

However, when it comes to vehicles, oil isn't the only kind of lubricant. Air and other gases can also be used, and this is how a hovercraft moves so easily once it's floating.

If a hovercraft is floating on a large enough cushion of air, it can go over bumps, up a slight incline, and even go over water.

Which raises an interesting question: Is this Leaf-Blower Hovercraft road-worthy or sea-worthy? Sadly, not really. Since this relatively simple-to-build hovercraft design creates an air cushion that is only about 1" to 2" thick , it can only travel easily over smooth surfaces such as a wood or cement floor or a very smooth road. For a hovercraft to be able to move easily over rougher terrain—like a gravel road, a lawn, or the rippled surface of a lake—you'd need a thicker air cushion because you'd need to be higher above the surface of whatever you were traveling over. You'd need to make sure that none of the pebbles, rocks, plants, ripply waves, or just bumps in the road interfere with the bottom of the hovercraft as it slides along.

Hovercrafts that are designed to travel over rough roads or other uneven terrain create an air cushion that is usually 6" to 9" high, with a skirt around the base of the hovercraft that contains the air.

If you want to build a hovercraft that you can drive on roads and over water, all you need to power it is an old lawn mower engine. Check out EepyBird.com for details!

For more ideas, videos, and variations, visit www.eepybird.com/experiments/hovercraft.

In this experiment, fifteen simple pendulums, each slightly longer than the next, create a mesmerizing display almost like a kaleidoscope. When you start them all swinging, they swing together. But as they continue to swing, different pendulums synchronize with each other while others fall out of phase. At one point, every third pendulum synchronizes, so you can see three distinct "waves" in the swinging pendulums. After a few more seconds, every other pendulum is synchronized, so you can see two waves moving through each other. And finally, amazingly, after a minute or so, all the pendulums come back into phase with each other, as the machine returns to one synchronized wave.

HOW DOES IT WORK?

The Pendulum Wave Machine is based on the principle that the period of a pendulum (the time it takes to swing) is determined by the pendulum's length. It doesn't matter how heavy or light the weight on the end is. The longer the pendulum, the longer it will take to swing back and forth.

In this experiment, because each pendulum is slightly longer than the next, each pendulum takes a slightly longer time to swing back and forth than the one after it. That means that, although they start swinging together, the shortest pendulum starts its return swing just a moment before the pendulum next to it, which starts its return swing just a moment before the one next to it, and so on all the way down the line. This is what creates the initial snaking wave motion.

Eventually that first wave is no longer discernible, and after a few moments of what looks like chaos, a new pattern emerges, all because each pendulum is moving slightly faster than the one next to it.

The pendulums in Pendulum Wave Machines are adjusted to such precise lengths that one pendulum will complete, say, 50 swings in 60 seconds, while the next one will complete 49 swings in 60 seconds, and

the next will complete 48, and so on. That means that at the end of the full time period (in this example, 60 seconds), all the pendulums will have completed these precise numbers of swings, they will all be over on the same side together, and they will all be synchronized again for a moment.

Why, you wonder, do multiple wave patterns occur? Halfway through the full time period, the first pendulum will have completed 25 swings, whereas the one next to it will have completed 24½. At that point, they'll be exactly opposite each other, and at that moment, two distinct waves will emerge: all the "even" pendulums will be on one side, all the "odd" pendulums on the other, and you'll be able to clearly see these two waves moving through each other.

Similarly, when a third of the full cycle is complete, every third pendulum will be synchronized for a moment, and you'll see three waves moving through each other. If you're sharp eyed, you'll be able to see the quarter mark of the complete cycle when every fourth pendulum is synchronized.

The Pendulum Wave Machine is an extraordinarily simple concept that takes a high degree of precision in order to work well. For these pendulums to really do their stuff—and it's truly hypnotizing when they do—they need to be carefully adjusted so that the lengths are just right. But fear not, we've got a simple design that will make adjusting the exact length of your pendulums easy to do.

The initial tuning may take as much as an hour or so, but once that work is done, keeping your pendulums in shape is a breeze.

To adjust the pendulum lengths precisely, this design borrows the time-honored tuning peg system employed in string instruments for thousands of years. Even the Lyres of Ur, a collection of ancient harp-like stringed instruments that dates back forty-five hundred years, used an early version of the tuning peg.

Here, each pendulum hangs from a piece of fishing line that is attached at one end to a tuning peg. Turning the peg gently will allow you to decrease or increase the length of your pendulums to an accuracy of within a tiny fraction of an inch, which is what you need in order to get the beautiful and complex repeating wave patterns this machine can create.

You'll be building a series of 15 separate pendulums each slightly longer than the next. Each pendulum is attached to the board at two points, which insures that it can only swing in one plane. Attached at a single point, a pendulum can move in any direction, but a pendulum attached at two can swing in only one direction.

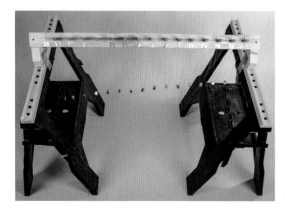

MATERIALS

- one 42" length of 1" x 3" wood strapping
- two 3' lengths of ½" wooden dowel
- masking tape
- one 6½" length of 2 x 4 lumber
- two 1¼" drywall screws
- fishing line
- duct tape
- 15 hexagonal nuts approx. ½" diameter
- one wide board, approximately 38" x 4" (we use a strip of ½" plywood that's 38" x 6")

TOOLS

- handsaw or chop saw
- yardstick
- drill, with ½" drill bit (or use a drill press, which is recommended)
- sandpaper
- 2 sawhorses, tables, or chairs (of equal height)
- screw gun or screwdriver
- scissors
- measuring tape

HOW TO BUILD IT

Building the Spine

The first step is to make the spine—the board with the tuning pegs from which your pendulums will hang.

STEP ONE: Measure and cut, using the saw, the 42" piece of strapping. Make a pencil mark 3" in from each end, and at each mark, draw a line across the width of the strapping.

STEP TWO: The yardstick (which is 3' or 36" long) should fit between these two lines along the length of

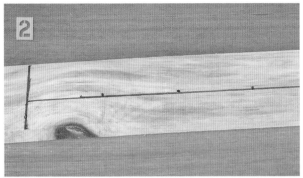

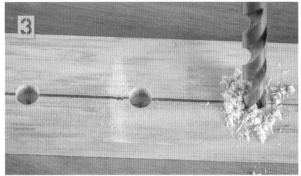

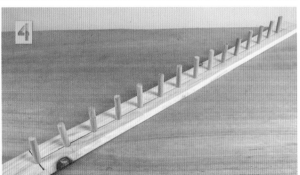

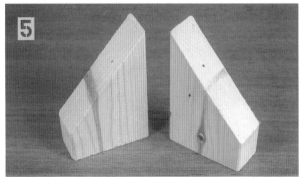

the board. Place the yardstick between the lines, flush with the bottom edge of the board. Starting at one end, mark the board directly above the yardstick every 2½". You'll end up marking 16 spots, all 2½" apart and a yardstick's width from the bottom edge of the board. The first 15 of these spots will be the locations of the holes for your tuning pegs.

NOTE: The ends of board (the initial 3" segments marked in STEP ONE) should remain clear. This is where you will mount the pendulums for display once everything's been assembled.

STEP THREE: Using the drill and the ½" drill bit, drill 15 holes, one centered at each of your first 15 marks. These holes don't have to go all the way through. Drill just half or three-quarters of the way through the board, but don't worry if you drill all the way through (so long as you don't drill through into something important).

STEP FOUR: Using the saw, cut fifteen 2½" lengths of ½" dowel (out of the two 3' lengths) and insert them into

the newly drilled holes. These are the tuning pegs. The dowels want to fit snugly into the holes but not so tightly that you can't turn them. If they are too snug and won't turn at all, sand them down with a piece of sandpaper. If they are too loose, wrap a little bit of masking tape around the end of the dowel to increase the thickness until they fit snugly.

STEP FIVE: Now, cut 2 legs to mount the spine on. Using the chop saw (preferably), cut one 6½" length of 2 x 4 lumber; make the cuts as straight as you can so that the legs won't wobble. Then you will cut this 2 x 4 in half at an angle to make two identically angled mounts. From one end of the 6½" block, and along one edge, make a mark ½" down. Then, from the other end of the block, and along the opposite edge, make a mark ½" down. Draw a line connecting these two marks across the width of the block, so you have a nice diagonal line that starts ½" from each end. Cut along this diagonal, again trying to get a nice straight cut. Set these 2 identical legs on end, with the point up.

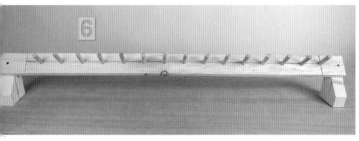

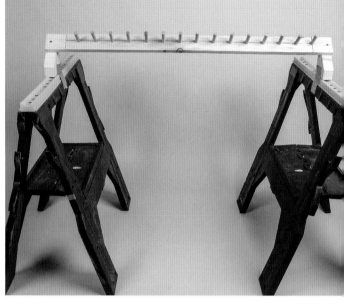

STEP SIX: Attach the legs to each end of the spine, so that the entire spine sits at an angle. Set the legs upright, 42" apart, with the short side closest to you and the diagonal cut running up and away from you. Place the spine on top of the 2 legs, with the top edge of the spine flush with the raised points of the legs and the entire spine tilted toward you. Using the screw gun and 1 screw, attach each end of the spine to the diagonal of the leg.

Hanging the Pendulums

Place 2 sawhorses that are the same height (at least 18" tall, but higher is nice) about 3' apart and set the spine so that it spans the gap between them. It's a good idea to attach the legs of the spine to the sawhorses with a little duct tape to help avoid accidentally knocking the whole thing over.

If you don't have sawhorses, you can use 2 same-height chairs or tables. Ideally, you want easy access both from the front and from the side, so you can adjust the tuning pegs and then go over to the best viewing position: looking along the length of the spine at the pendulums swinging underneath.

Get comfortable, because hanging and tuning your pendulums to the exact right length can take a while. You may want to get a chair to sit on for yourself.

STEP ONE: Using the scissors, cut a 3' length of fishing line.

STEP TWO: Starting at one end of the spine, use a little piece of duct tape to tape one end of the piece of fishing line to the side of the first tuning peg. (You're going to need a lot of little pieces of duct tape—30, to be precise—so you may want to make a bunch of little strips and have them standing by.) Wrap the fishing line around the peg a few times and make sure the tape is thoroughly stuck to the peg so that the fishing line won't pull away.

STEP THREE: Thread the other end of the fishing line through one of the hex nuts and then bring the end back up to the spine. With another little piece of duct tape, tape the loose end of the fishing line securely to the face of the board, directly below the next tuning peg. This is your first pendulum. It should swing nicely back and forth from its two pivot points where the fishing line hits the edge of the spine.

STEP FOUR: Repeat STEPS ONE through THREE for the other 14 pendulums. Cut a 3' length of fishing line, attach it to the next tuning peg, wrap the line a few times around the peg, thread a hex nut, and tape the loose end onto the face of the board beneath the following peg. You'll end up with 15 hex nuts hanging on 15 Vs of fishing line.

NOTE: For the final or 15th hex nut, tape the loose end of the fishing line onto the board at a point roughly where the next tuning peg would be if there were one more peg.

STEP FIVE: Using a saw, cut your wide board to approximately 38" by 4". What's important here is the length. If your board is wider (say, 6") that will make it a little easier to use, but the width isn't crucial. This is your *swing board*, and you use it to get all the pendulums swinging together.

STEP SIX: Practice swinging all the pendulums at once: hold the swing board by both ends, place the board lengthwise behind all the pendulums, and scoop the board back toward you so that it pulls all the pendulums a few inches toward you. Keep the face of the swing board almost vertical, and make sure all the fishing lines stay taught. Release the pendulums by pulling the swing board down quickly. This starts them all swinging simultaneously.

NOTE: You don't *hit* the pendulums to get them to swing; you gently pull them toward you and release them. Also, you don't need to get the pendulums swinging very hard, so don't pull them high up. A small swing is better than a big swing. Pendulum behavior changes a bit the harder they swing. So just pull them toward you a few inches and let them swing smoothly and simultaneously.

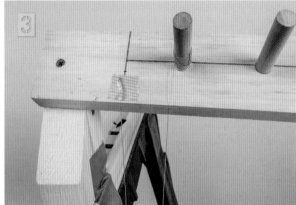

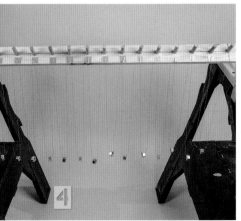

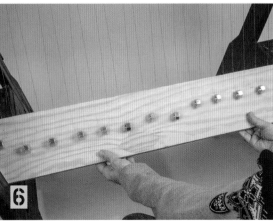

Tuning the Pendulums

It's time to precisely adjust the lengths of the pendulums. The only tools you'll need for this are the measuring tape and lots of patience.

STEP ONE: Using the measurements below, start by getting the lengths of the pendulums approximately correct. Measure from where the fishing line dangles over the edge of the spine to the center of the hex nut (not the top or bottom, but the center). The measurements for all 15 pendulums should be as follows:

→ 13½"   → 11⅝"   → 10⅛"   → 8⅞"

→ 13"   → 11¼"   → 9¾"   → 8⁹⁄₁₆"

→ 12½"   → 10¹³⁄₁₆"   → 9⁷⁄₁₆"   → 8⁵⁄₁₆"

→ 12¹⁄₁₆"   → 10⁷⁄₁₆"   → 9⅛"

These measurements are all approximate. The Pendulum Wave Machine will need to be far more precise than this to work properly, but you start by getting close to the correct measurements and then you fine tune from here.

One at a time, from one end to the other, adjust each pendulum as close to these lengths as possible. Depending on how far off each hex nut is, first wrap the fishing line around the tuning peg enough times to raise it up to almost the exact height, and then twist the tuning peg to make fine adjustments. Once you have one pendulum close to the correct length, move onto the next.

NOTE: As you adjust the lengths of the pendulums, it's important that the fishing line be wrapped around the base of each tuning peg, right next to the board of the spine. If you coil the fishing line high up on the peg, the pendulum won't swing properly, so make sure you keep the coil all the way down at the bottom of the peg.

STEP TWO: Now tune them exactly. Start with just the first 6 pendulums. Take the swing board, pull back the 6 longest pendulums together, and release them at the same time. Again, don't pull them too far back. A short swing makes the pattern easier to see.

Watch from the end of the spine on the side with the longest pendulum. You're looking for a clean snaking wave pattern in the first 5 to 10 swings. After that there will seem to be no pattern for a few seconds, and then you should see 5 to 10 swings of the following (with a few seconds of chaos in between each):

→ 3 pairs of pendulums swinging together (3 waves)

→ 2 trios of pendulums swinging together (2 waves)

→ 3 pairs of pendulums swinging together again

→ 1 snaking wave again

You will likely need several test swings to see which pendulums are out of synch. This is why testing the first 6 pendulums is best, since fewer pendulums are less confusing to follow. To start, you should be able to pick out one or two that are off. This is easiest to see as the

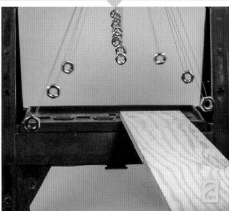

TWO DISTINCT WAVES EMERGING WITH 6 PENDULUMS SWINGING.

cycle begins to repeat itself, and you can see the single snaking wave for the second time. Ideally, this single wave should look exactly like the very first wave that you get right after you start the pendulums swinging.

Eyeball the wave, identify the problem pendulums, and make small adjustments by turning the pegs to lengthen or shorten the pendulum lengths. The shorter the pendulum string, the faster the pendulum will swing, so if a pendulum seems to be a little ahead of the pack, slow it down by lengthening the string a touch. If it seems to be behind, speed it up by shortening the string. Tighten or loosen the strings in very small increments. Very small adjustments will make a difference.

If it isn't clear whether a pendulum needs to speed up or slow down, make a guess and give the pendulums a swing. You'll see quickly if you've made things better or worse.

STEP THREE: Once the first 6 pendulums have been tuned to the right length, add 1 or 2 pendulums at a time and check the group pattern each time. Swing the 7 longest pendulums together and tune the 7th to precisely the right length. Then do 8, 9, and so on.

The final product should produce these recognizable patterns:

→ 1 wave

→ 4 waves

→ 3 waves

→ 2 waves

→ 3 waves

→ 4 waves

Then the wave pattern will repeat. If you color-coded the hex nuts, you'd be able to see even more wave patterns, but it's pretty easy to see up to four waves without adding color.

It's also important to note that viewed from the side, the hex nuts at the bottom of your pendulums will *not* line up in a straight line, but instead will form a slight curve. Once you can see the slightly curved line formed by the bottom of the pendulums, you can also use that to help your tuning.

THE SCIENCE

Scientific exploration of the properties of pendulums started with a teenage boy sitting in church over four hundred years ago.

Sitting on a pew in a church in Italy in the year 1582, so the story goes, seventeen-year-old Galileo Galilei's mind began to wander as he watched a chandelier move back and forth ever so slightly in the air currents.

The young Galileo noticed that the chandelier seemed to take about the same amount of time to complete its swing whether it was moving a lot or a little. With no clock or watch to time the swings, he used his own pulse to measure time and confirmed his suspicions. A wide swing and a short swing both took the same amount of time.

Years later, Galileo—the father of modern science—did more precise experiments and established the principal of *isochronism*, the property of a pendulum to maintain a steady and consistent beat regardless of the degree to which it swings. This became the foundation for modern clocks and timekeeping—and the Pendulum Wave Machine.

Pendulums are remarkable devices because they are so consistent and predictable. This steady regularity, no matter how wide the arc of the swing, is what keeps the pendulum in a grandfather clock so predictable, gaining or losing very little time as it runs.

The period of any given pendulum is determined by its length and not, as you might suspect, by its weight. As we've seen with the Pendulum Wave Machine, the

THE FORCE OF GRAVITY: THE LAW IS DIFFERENT IN DIFFERENT PLACES

We're fond of the humorous bumper sticker: "Gravity: It's Not Just a Good Idea—It's the Law!" While that's true, you might be surprised to learn that the "law" of gravity isn't the same everywhere. More precisely, the "law" exists everywhere, but we experience it slightly differently in different places. This is because the Earth isn't a perfect sphere and it isn't holding still. Because the Earth is covered with mountains and valleys, because it's a little wider across the middle than from top to bottom, and because it's constantly spinning at about a thousand miles an hour (at the equator), the degree to which gravity affects us changes depending on where we are.

Think of the Earth like a merry-go-round. As on a merry-go-round, how fast you're moving while you're on it depends on how far away from the center you are. At the poles, just as in the center of a merry-go-round, you barely move at all. At the equator, you're as far from the axis around which the Earth is rotating as you can get (without leaving the ground), so that's where you're spinning the fastest. So thanks to what's called centripetal force

(the force that wants to send you flying off the merry-go-round as you get closer to the edge), the closer you are to the equator, the less you weigh. At the North or South Pole, you'll weigh about half a percent more than you do at the equator.

Also, the farther you are away from the Earth's center of gravity, the less it affects you. On Mount Everest, you weigh about a quarter percent less than you do at sea level. Hard to believe as it may be, the thickness of the atmosphere also factors into how we experience gravity. The thicker atmosphere we experience at sea level gives us a touch more buoyancy than the thinner air we find at higher elevations.

All of that, combined with the little bit of elasticity of the fishing line and the inconsistency found in most lumber, means that you'll need to fine-tune your pendulums to the lengths that work best for you once you've built them. Your Pendulum Wave Machine will behave slightly differently at sea level than it will on top of Mount Everest. The tuning peg mechanism makes this pretty straightforward to adjust for.

longer the pendulum, the longer it will take to swing back and forth. If you know the length of a pendulum, you can figure out quite precisely and quite easily how long it will take to swing back and forth.

This is slightly less true for wide swings, but for pendulums like those in a typical grandfather clock—or for those in this experiment—the period remains essentially the same whether the pendulum is swinging at its maximum or just barely moving. As the swings get wide, the period becomes just a little longer. For example, when a pendulum's swing reaches 23 degrees, its period is about 1 percent longer than when it is barely swinging. That's one reason we say not to swing the Pendulum Wave Machine too hard.

For more ideas, videos, and variations, visit www.eepybird.com/experiments/pendulumwave.

HOW TO CALCULATE THE PERIOD OF A PENDULUM

Where did the list of lengths for the 15 pendulums in the machine come from? You can calculate the period of time a pendulum takes to complete one full swing back and forth with a simple formula. Armed with that formula and some algebra, you can figure out what length to make a pendulum if you want it to complete a certain number of swings in a certain period of time. We won't go through all that now, but here's a look at that first formula.

To calculate the period of time (abbreviated "T") that it takes for a pendulum to complete one full swing back and forth, you use the formula:

$$T \approx 2\pi \sqrt{L/g}$$

where L is the length of the pendulum and g is the force of gravity, which on Earth is around 32 feet per second squared. The formula is slightly approximate, because there's a tiny adjustment needed when the swings get larger.

A pendulum with a length of 5 feet then will have a period we can calculate like this:

$$T \approx 2\pi \sqrt{5 \text{ feet} \div 32 \text{ feet per second}^2}$$
$$\approx 2\pi \sqrt{0.156 \text{ second}^2}$$
$$\approx 2\pi \times 0.39 \text{ seconds}$$
$$\approx 2 \times 3.14 \times 0.39 \text{ seconds}$$
$$\approx \textbf{2.48 seconds}.$$

Similarly, a pendulum with a length of 2 feet will have a period calculated like this:

$$T \approx 2\pi \sqrt{2 \text{ feet} \div 32 \text{ feet per second}^2}$$
$$\approx 2\pi \sqrt{0.0625 \text{ second}^2}$$
$$\approx 2\pi \times 0.25 \text{ seconds}$$
$$\approx 2 \times 3.14 \times .25 \text{ seconds}$$
$$\approx \textbf{1.57 seconds}.$$

So based on these calculations, we can see that a 5-foot pendulum takes just over a second more time to complete a swing than a 2-foot pendulum does.

How do you capture the power of a Coke and Mentos Geyser? It's easy! And once you harness that power, you can send a minirocket car down the road over 100 feet.

Everything you need to build your own minirocket car can be found at your local hardware store, Home Depot, or Lowe's. The total cost should be less than twenty dollars, and the project only requires simple tools. That said, consider buying extras of all the supplies. Not only will you want to race the car multiple times, but nothing is more fun than making several different cars and racing them against each other!

HOW DOES IT WORK?

The nucleation reaction that powers this minirocket car is explained in the Coke and Mentos Geysers experiment on page 70. So, how do we harness that power to create propulsion?

When the Mentos and soda react, the carbon dioxide that is under pressure in the soda is released and expands. That expansion makes the soda shoot out of the mouth of the bottle. Stand the bottle upright and you get a geyser. Place the bottle on its side and all that energy can push your minirocket car down the road.

Still, it isn't enough just to let the soda shoot out by itself. When we were originally designing this car, that's actually the first thing we tried. We built something like this:

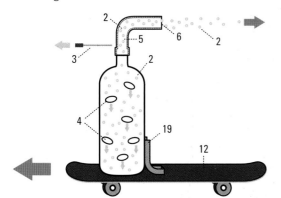

And it worked. But not very well. The soda shot out the back and the skateboard rolled forward, but it didn't go very far or very fast. Hmm, we thought: something's missing. What was happening was that most of the energy in the reaction was going toward spraying the soda out of the bottle and pushing against the air; it wasn't maximized to move the vehicle forward. If your goal is to move forward, every bit of energy that's spent on anything else is wasted.

So, for this experiment, we developed a simple piston-and-cylinder design that focuses almost all of the energy from the reaction on moving the car forward. When the reaction starts, the only place for the expanding CO_2 to go is into a PVC tube, which at first is almost entirely blocked by a wooden piston.

While you're holding both the piston and the cylinder together, nothing can move, and the pressure of the reaction just continues to build. That pressure is trying to either push the piston out of the cylinder, or push the car away from the piston. Once you let go of the cylinder, all of the pent-up energy of the expanding CO_2 pushes against the piston and sends the cylinder, bottle, and wheels in the opposite direction. In a spray of soda-exhaust, the minirocket car zooms down the road.

THE EXPERIMENT: THE MINIROCKET CAR

First, you build, and then you launch. With this project, you definitely need to be outside and have a nice open, flat, wide, long space in front of you that's away from traffic and clear of people, pets, and any stationary objects that might stop your minirocket car prematurely. Don't underestimate the distance this might go, nor the fizzy mess it will create.

You can strap your Coke and Mentos engine to all sorts of vehicles. Just about any wheeled platform will work, such as a skateboard. We've found that the best simple set of wheels for this car is a tri-dolly. It's light, stable, and works really well. It's about 6" across, has three rotating casters, and costs about nine dollars.

TRI-DOLLY

MATERIALS

- one 30" piece of ¾" wood dowel
- one 4" piece of 1" x 3" wood strapping
- one 12" length of nylon ratchet strap, canvas, or other strong cloth
- one ¾" wood screw
- one 2" drywall screw
- one 24" length of ¾" PVC pipe
- masking tape
- 6 Mentos mints per launch (get several packs)
- one 2-liter bottle of diet soda per launch (room temperature, not cold!)
- duct tape
- tri-dolly (or skateboard or other wheeled platform)

TOOLS

- handsaw or chop saw (and/or a hack saw for cutting the PVC)
- scissors
- screw gun or screwdriver
- safety goggles
- lab coat (or raincoat)

HOW TO BUILD IT

STEP ONE: Make the piston (which is the key to harnessing the power of the Mentos geyser). First, using a saw, cut a 30" length of ¾" dowel; this is the piston rod. Then, cut a 4" piece of 1" x 3" wood strapping; this is the handle block.

STEP TWO: Using scissors, cut a 12" length of either nylon ratchet strap, canvas, or strong cloth. This is for the wrist strap. If your hands are large, measure a slightly longer strap.

STEP THREE: Fold the strap to make a loop. Using a screw gun and the ¾" screw, attach the open ends of the loop to the handle block as shown. Be careful screwing the strap down and take it slow, since the screw can grab the strap and try to wrap it around the screw.

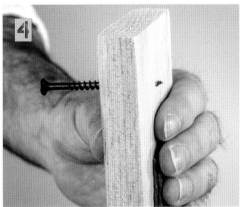

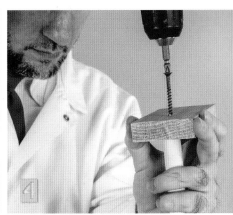

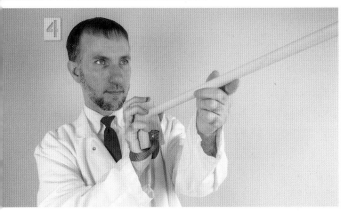

STEP FOUR: Attach the dowel. On the opposite side of the handle block, and above the wrist strap, screw the 2" screw into the block, but stop when the tip of the screw just pokes out the opposite side. (In the picture we do this on top of a block of scrap wood so that the screw doesn't accidentally go into our table.)

Now stand the dowel on end and place the wood block on top of it, carefully lining up the exposed tip of the 2" screw with the center of the end of the dowel. Continue screwing the rest of the way till the handle block is firmly attached to the dowel. This is tricky, and it can be good to have someone else help by holding onto the dowel while you put in the screw. Be careful to line up the screw straight up and down so that it doesn't pop out the side of the dowel!

STEP FIVE: Make the piston pressure chamber (and Mentos holder). Using a saw, cut a 24" length of ¾" PVC pipe.

Now you're done with all the tools. It's time to go outside and get ready to launch!

NOTE: We have a rule in our laboratory that no one is allowed to open a bottle of Coke inside the lab when Mentos are in the vicinity. The possibility of a big mess is too great (as is the temptation for mischief). Always head outside before you start opening the soda bottles!

Prelaunch Preparation

Before each launch, you prepare the same way.

STEP ONE: Place a strip of masking tape over one end of the PVC tube, sealing it. Although it won't be a perfect seal, it will cover most of the end of the tube.

STEP TWO: Drop 6 Mentos mints into the open end of the tube. The mints should slide to the bottom of the tube and rest on the masking tape seal at the bottom. Keep the tube upright so the Mentos don't fall out.

STEP THREE: Open the 2-liter bottle of room-temperature diet soda. For more on soda type and temperature, see the Coke and Mentos Geysers experiment, page 70. Remember: Don't use soda directly from the refrigerator. Cold soda will make for a very slow minirocket car! Place the masking-tape-sealed end of the tube (with the Mentos in it) just onto the mouth of the bottle.

Using duct tape, secure the tube tightly and solidly onto the mouth of the bottle.

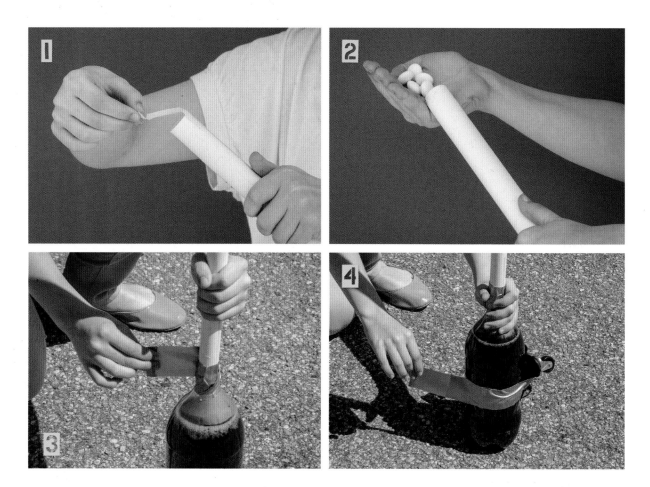

STEP FOUR: During this step, make sure to hold the bottle upright so no soda spills out. Now, tape the tridolly onto the side of the bottle with duct tape. Your minirocket car is now ready for launch!

Launching the Minirocket Car

Important: Read through the entire launch instructions first! Once you tip the bottle on its side, the soda will start reacting with the Mentos immediately. At that point, there won't be any more time to read these instructions!

STEP ONE: Put on safety goggles and a white lab coat—or perhaps a raincoat, since you will probably get sprayed by soda.

STEP TWO: Put on the wrist strap attached to the block and piston. This is very important: once the car launches, the piston could easily fly out of your grip and hit someone accidentally. Lab coats are optional, the wrist strap is not.

STEP THREE: Pick a direction to aim the rocket car, and make sure it is clear of people, cars, buildings, and other nearby objects for at least 100' to 200'. Depending on the temperature and your launch experience, the minirocket car can go anywhere from 30' to almost 200'.

STEP FOUR: With the bottle standing up and the tube vertical, gently insert the piston into the tube until the piston just rests on top of the Mentos—but don't push it all the way in yet.

3/4" dowel

PVC tube

Mentos

masking tape

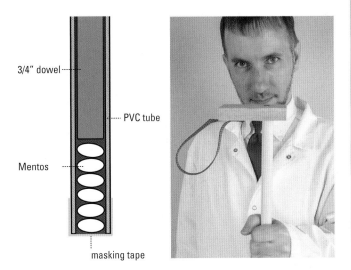

STEP FIVE: With one hand on the tube and the other on the piston handle, tip the rocket car down so the wheels are on the ground. As you tip the bottle, a little bit of the soda will begin to pour into the tube and react with the Mentos. This is okay, but don't let go yet!

STEP SIX: Once the car is resting on its wheels, immediately push the piston all the way into the tube so that the Mentos break through the tape and go into the bottle. *Keep holding the tube for leverage as you do this, and don't let the tube go once it's done.*

STEP SEVEN: Wait like this for one to two seconds. The soda and Mentos will react instantly once combined, and the bottle will start filling with foaming soda. The car will try to pull away, but don't let it. You want to let the reaction build; this is like revving the car's engine.

STEP EIGHT: After a few seconds, confirm you have a strong grip on the piston block (the wood strapped to your wrist), carefully aim the rocket car, push the whole assembly forward in a bowling motion, and when you're fully extended, let go of the tube. Prepare to get a little bit wet!

STEP NINE: Watch that rocket car go! And don't forget the most important part: celebrate!

How far can a single-bottle minirocket car go? It takes practice. Our first tests went 30'. Now that we've refined the methods, the farthest we've propelled a single bottle (so far!) is 176'. See what results you can get, and let us know how far your minirocket car goes!

This design is very similar in concept to the way the cylinder-and-piston system works in your car's engine. Just like in the minirocket car, an automobile engine harnesses the explosion of expanding gas in a cylinder in order to push a piston that turns the wheels and makes the car move. In a standard gasoline engine, a small amount of gasoline is sprayed into one end of the cylinder. A small electrical charge then causes the spark plug to spark, which ignites the gas vapor and causes a small explosion.

The exploding gas vapor pushes out in all directions, but there is only one direction the explosion can go: that is toward the piston, which is pushed away from the spark plug. Meanwhile, the piston is connected by a connecting rod (sometimes called a "con rod") to the crankshaft. Thus, as the piston moves, it turns the crankshaft, which turns the engine and ultimately the wheels of the car.

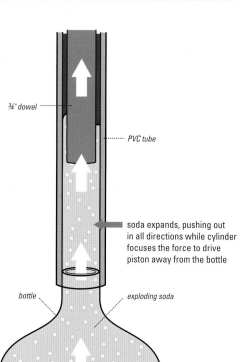

¾" dowel

PVC tube

soda expands, pushing out in all directions while cylinder focuses the force to drive piston away from the bottle

bottle

exploding soda

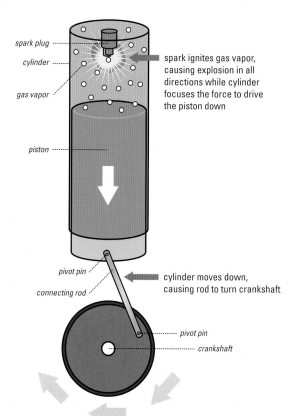

spark plug

cylinder

gas vapor

spark ignites gas vapor, causing explosion in all directions while cylinder focuses the force to drive the piston down

piston

pivot pin

connecting rod

cylinder moves down, causing rod to turn crankshaft

pivot pin

crankshaft

OUR 108-BOTTLE COKE-AND-MENTOS-POWERED ROCKET CAR IN ACTION.

Just like in a Coke-and-Mentos-fueled minirocket car, it's the cylinder-and-piston mechanism that focuses all the energy of the explosion to push the vehicle forward.

This Coke-and-Mentos-powered minirocket car, however, just uses one bottle. If you're like us, your reaction to doing this experiment will be: okay, now let's try that again with *more* soda bottles and *more* Mentos!

Our largest Coke-and-Mentos-powered rocket car (so far) is capable of carrying a human: it's powered by 108 bottles of soda and 648 Mentos, and it uses a similar piston mechanism to harness all that power. It has carried Fritz down the road 221 feet, and it carried talk-show host David Letterman over 360 feet—but

Dave had a bit of a downhill slope assisting him. How big can it get? We haven't found the limits yet!

For more ideas, videos, and variations, visit www.eepybird.com/experiments/rocketcar.

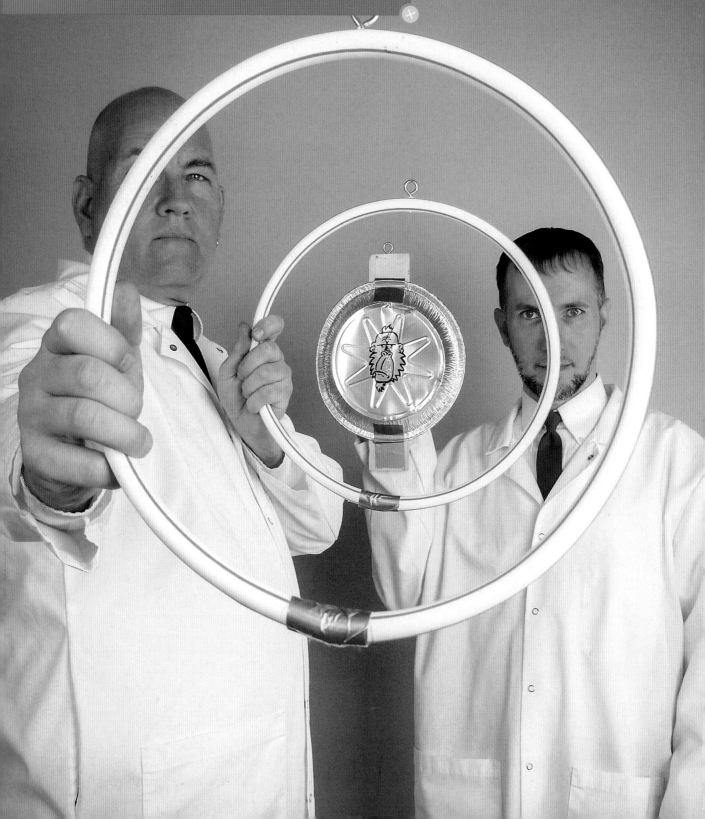

Imagine shooting a dart with a blowgun through two hoops and hitting a target. Now imagine that both hoops are moving—and the target is moving, too. Impossible! No. With the right setup, it works automatically. It's all physics in action.

This is our variation on a classic science demonstration called "Shoot the Monkey." Suppose a monkey is trying to evade a hunter's dart. The monkey is hanging in a tree, and the moment the gun fires, he lets go and drops from the tree. Where should the hunter aim so that the dart will still hit this moving target?

The surprising answer is that the hunter should aim directly at the monkey.

HOW DOES IT WORK?

To understand how a hunter can hit a falling monkey this way, let's simplify the situation slightly and imagine that the monkey is at the same height as the hunter's head. If the hunter aims straight ahead, what will happen?

Gravity acts on the dart as it flies, so the dart will be pulled down toward the ground in an arc. If the monkey lets go and starts falling the instant the dart is fired, he will fall straight down. So here's the key question: which falls faster, the dart or the monkey?

Since many light objects like feathers and leaves fall slowly, and many heavier objects like sticks and stones fall more quickly, you might think that the heavier object, the monkey, would fall faster. That's what the ancient Greek philosopher Aristotle thought. Well, he probably didn't think in terms of monkeys and darts, but he thought that a heavy object would fall faster than a light object.

As we'll see, however, Aristotle was wrong. This experiment demonstrates that gravity does indeed have the same effect on both the monkey and the dart.

For this experiment, you'll need a big, open space, like a backyard, a garage, or a gym. In addition, this project involves more construction than other experiments, and it has a lot of pieces and materials to buy. However, it's not difficult to build if you're familiar with basic carpentry.

Because it's slightly more complex, we've broken the description into four main sections: building the drop structure, making the dart gun, making the dropping mechanism (and the monkey), and then conducting the experiment.

The drop structure, which holds the two hoops and the monkey, is essentially a large, freestanding doorframe that you will construct. You'll need one or two friends to help make this.

Figuring out how to make a simple, nonexplosive gun or a crossbow that accurately shoots a projectile was difficult. We went through a lot of different designs before we came up with the blowgun here, which is simple and surprisingly consistent.

Then, instead of using a real monkey (which would be unkind, not to mention illegal), you make a pie-tin monkey and a mechanism for dropping it, and then you're ready to go.

TOOLS

- handsaw, chop saw, or jig saw
- screw gun
- square
- ladder
- 3 spring clamps (each big enough to clamp 2 boards together)
- scissors
- markers (to draw on the pie tin)

MATERIALS

The drop structure

- ten 8' lengths of 1" x 3" strapping (this should be enough for all parts of the experiment)
- a box of 1¼" drywall screws
- 6 eye hooks (size 10 works well)

The blowgun

- duct tape
- one 18" length of flexible ⅝"-diameter tubing
- 1 empty 2-liter soda bottle
- one 2' length of ¾" PVC pipe
- one 18" length of 1" x 3" strapping
- one 9" length of 1" x 3" strapping
- one 12" length of 1" x 3" strapping
- two 1' x 2' pieces of ½"-thick plywood
- one 4" x 5" piece of craft foam (available in 9" x 12" sheets at hobby and craft stores)
- surveyor's tape (or a similar, very lightweight ribbon)

The drop trigger

- 1 hula hoop (roughly 25 to 30" diameter)
- two 2" long ½" diameter dowels
- 5 eye hooks (size 10, or the same size as for the drop frame)
- 1 pie tin
- one 18" piece of 1" x 3" strapping
- one 12" piece of 1" x 3" strapping
- scraps of lumber of varying thickness (a 6" tall stack of ½-1" thick boards, all around 1-2' long)
- string
- 3 cotter pins
- chair (one where the top of the seat back is straight, not curved)
- one 3' length of ¾" dowel

HOW TO BUILD IT

The Drop Structure///////////////////////////////////

STEP ONE: Using a saw, cut the 10 lengths of strapping to the following measures. Make your cuts as straight as possible to help ensure that your frame will be square. A chop saw is designed to make perfectly square cuts.

→ Two 8' lengths → One 18" length

→ One 5' length → Two 9" lengths

→ Two 4' lengths → One 6" length

→ Three 3' lengths → Three 3" pieces

→ One 30½" length

NOTE: The first 2 pieces of 8' board will be the upright legs of your frame. If your plan is to build and use this indoors, consider the height of your ceiling. If your ceiling is at or less than 8' tall, make these legs shorter, so they will fit, but as tall as the ceiling allows.

STEP TWO: Assemble the 2 legs that will support the frame, using the two 8' lengths, the two 4' lengths, and two of the 3' lengths.

Lay the two 8' lengths flat on the floor. Using the screw gun and 2 screws for each, attach one 4' length flat across the end of each leg to make a T shape (see photo). Use the square to ensure each T is square, so that your frame will stand up well—each T will end up upside down with a top crosspiece connecting them.

Next, brace each T with a 3' length of strapping. With each T flat on the floor, lay a 3' length diagonally and flat from the upright to one end of the 4' length. Attach this diagonal brace with a screw on each end, making sure the bottom corner of the brace doesn't extend below the bottom of the 4' length (which needs to be flush to the floor for stability).

STEP THREE: Using a ladder (and preferably help from one or two other people), complete the frame. Stand the 2 T structures upside down (and reverse the orientation of their braces for added stability), then place the third remaining 3' length flush and flat across the tops of the uprights. From the top, screw 2 screws into each end of the 3' crosspiece.

To further brace the frame, use screws to attach the two 9" pieces as diagonal braces to the top corners of the frame (see illustration). This completes the frame structure.

STEP FOUR: Build the E-shaped target support. This will hang face down from the frame's crosspiece.

On the 5' length of strapping, mark the

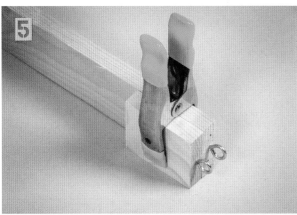

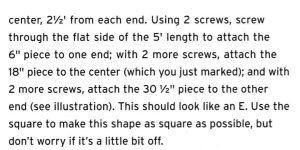

center, 2½' from each end. Using 2 screws, screw through the flat side of the 5' length to attach the 6" piece to one end; with 2 more screws, attach the 18" piece to the center (which you just marked); and with 2 more screws, attach the 30½" piece to the other end (see illustration). This should look like an E. Use the square to make this shape as square as possible, but don't worry if it's a little bit off.

STEP FIVE: Now attach the 6 eye hooks. Screw 1 eye hook into the end of each of the 3 prongs of the E, and screw 1 eye hook into the end of each of the three 3" pieces of strapping. Orient the eye hooks so they are parallel to the faces of the boards. Using the spring clamps, clamp a 3" piece onto the end of each of the 3 prongs of the E; align the eye hooks at the end of each prong so that you can look right through each pair.

STEP SIX: Now let's put it together. Once again, use the

ladder and some extra people to do this safely. Orient the E so that the prongs point toward the floor, position it perpendicular to the top crosspiece of the frame, with the center of the E roughly lined up underneath the center of the top crosspiece. Use 2 screws to screw down through the center of the top crosspiece into the E to attach it in place. Take a look at the photo above to see what the finished drop structure looks like.

The Blowgun

STEP ONE: Wrap duct tape around the end of the 18" piece of flexible tubing, building up layers of tape so that the taped end of the tube will fit tightly into the mouth of the soda bottle, while still keeping the airway clear. Wedge the taped end of the flexible tubing just inside the mouth of the bottle, and wrap duct tape around the mouth and tube to attach the tube securely to the bottle. Insert the other end of the flexible tube about 1"

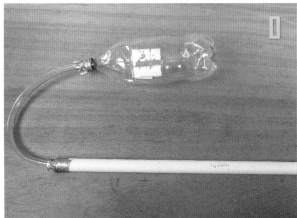

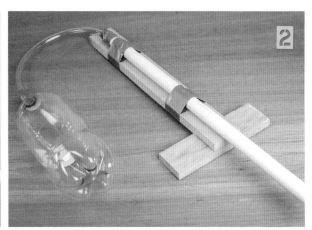

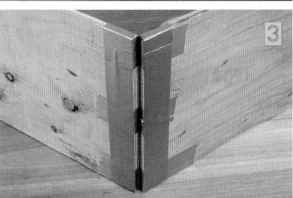

into one end of the 2' piece of ¾" PVC pipe. Wrap duct tape around the tube/PVC connection to make it tight.

STEP TWO: Using a saw, cut one 18" length and one 9" length of strapping. Laying the wood flat on the floor, make a T shape with the 9" piece laying flat across one end of the 18" piece. Use 2 screws to attach them. Flip the T over so that the 9" crosspiece is now underneath the 18" piece. Lay the PVC tube along the 18" piece of the T, so that the base of the T extends a few inches below the junction where flexible tube goes into the pipe, and so that the other end of the pipe extends past the 9" crosspiece. Wrap duct tape around the pipe and the 18" piece to hold them tightly together. This completes the blowgun.

STEP THREE: Now make a foot-operated stomper to generate air power. Using the jig saw, cut two 1' x 2' pieces of ½"-thick plywood.

Lay them on top of each other, aligning their dimensions. Choose one 1' end to be the hinge. Run several strips of duct tape over the outside of that edge, where the boards meet. Open the hinge and run strips of duct tape along the inside as well. Finally, to make your hinge strong, wrap a strip of duct tape completely around each piece of plywood, covering the duct tape used to make the hinge (see illustration). This will help keep the duct tape from pulling away from the wood.

Using a saw cut a 12" piece of strapping. Open the stomper and on the front bottom edge opposite the hinge, duct tape the piece of strapping flat along the 1' edge of the plywood. This strapping will function as a stopper, holding the bottle in place and keeping it from being squeezed out when you stomp. Now insert the bottle into the jaws of your stomper, parallel to the strapping, as shown in the photo above.

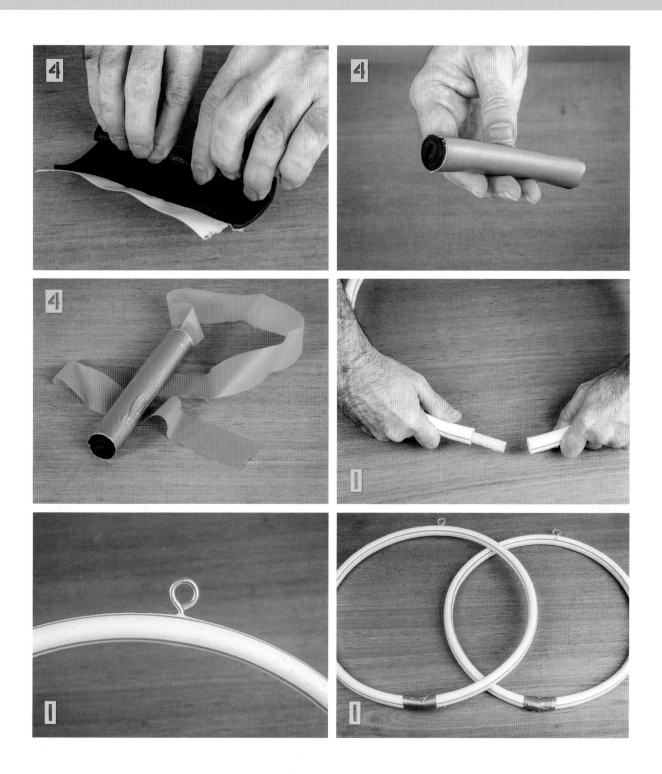

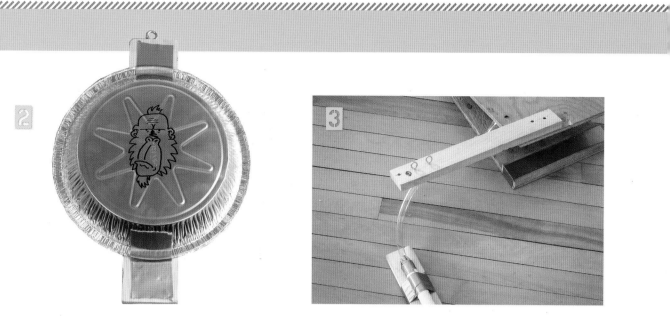

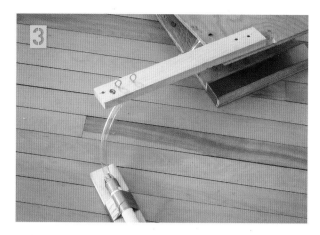

STEP FOUR: Make the dart. Using scissors, cut the 4" x 5" piece of craft foam. Put a 5" strip of duct tape along one 5" side of the craft foam with the edge of the foam running along the middle of the duct tape. Place the foam on the table with the sticky side of the duct tape up, and starting with the edge of the foam opposite the tape, roll the foam as tightly as you can toward the duct tape. Roll it right onto the duct tape so that the tape wraps around it to give you a nice tight cylinder of foam. Check the diameter of the dart by partially inserting the ends of the dart into the PVC pipe on your blowgun. Make sure that both ends of the dart slide easily in and out of the pipe. If the dart is too big, undo the tape and roll it tighter.

Duct tape a little bit of surveyor's tape (or other similar, very lightweight ribbon) to the end of the cylinder; now you have a dart with a tail to give it some drag, which helps it fly straight through the air. Don't use much duct tape or the dart won't slide smoothly inside the PVC pipe.

WARNING!

You have now built a stomp rocket/blowgun that can propel a projectile quite far and quite fast. Don't point it at anyone. Don't shoot it at anything breakable. It's only air and a piece of foam, but it can hurt and do real damage. Be careful!

The Drop Trigger and Monkey Target

STEP ONE: Start by turning the hula hoop into 2 smaller hoops. Using a handsaw, cut the hula hoop in half. Bend each half into a smaller hoop, inserting the 2" pieces of ½" dowel into the open ends to help join the ends together. Duct tape around the joint to keep the hoop closed. After both rings are complete, carefully screw an eyehook into the outer edge of each ring (it's easiest to screw into the plastic, not the dowel).

STEP TWO: It's monkey time. Using markers, decorate your tin with a monkey face, in honor of the original name of the experiment, or in any way you'd like. Once it's done, screw an eye hook into the end of a 12" piece of strapping; center the eye hook across the board's width. Position the pie tin along the middle of the flat side of the strapping and tape it down. Your target is complete.

STEP THREE: With 2 screws, attach the 18" piece of strapping to the top right-hand corner of your stomper, parallel to the 1' edge and overhanging off the side by about 12". Screw 2 eyehooks into the top, flat side of the board, near the end, parallel, and about 2" apart.

STEP FOUR: Set up your blowgun on the floor so that the end of the barrel is about 10½' away from the first and lowest drop point on the target support "E" attached to the drop structure. Place the chair over the PVC pipe so that the pipe sticks out the front and the flexible tube runs out the back. The back of the chair

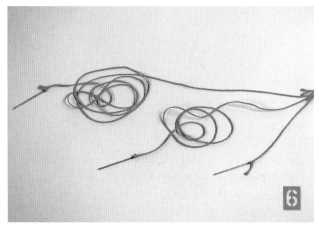

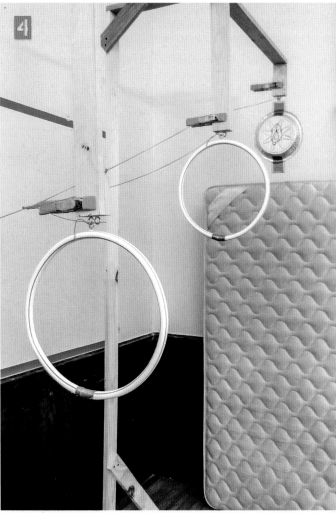

should have a straight top so when you run string over it, the string won't slide off. Place the stomper behind the chair so that the 12" overhanging piece of strapping is directly behind the center of the chair. It should look like the photo shown on the next page.

When everything looks good, tape the stomper and the back end of the blowgun's T-frame to the floor (make sure that the tape won't harm your floor). Put a few pieces of scrap wood underneath the front of the T to raise the barrel of the blowgun. Later, when you're ready to fire, you'll calibrate the height of the blowgun barrel more carefully.

STEP FIVE: How do you get the targets to drop when the blowgun fires? Fancy versions of this experiment use lasers and electromagnets. Our DIY version: string. The drop is not instantaneous, but it still works.

Cut 4 lengths of string as follows: 18', 1', 3½', and 6'.

With each of the 3 shorter strings, tie a cotter pin onto one end. These pins will hold the targets in place.

STEP SIX: Lay the 18' string on the floor and tie the loose ends of the other 3 strings to one end, attaching all of them to the same spot. It should look like the photo shown on page 176.

Ready, Aim, Fire ////////////////////////////////////

Now it's time to connect everything, calibrate the blowgun, and see this experiment in action!

STEP ONE: Take the loose end of the long string and wrap it loosely around the eye hooks on the stomper (in a figure-eight pattern). This is just to hold the end in place while you set the cotter pins. Run the rest of the string over the back of the chair and to the drop structure.

STEP TWO: Start with the target support drop point closest to the blowgun, on the end of the longest vertical piece of the E. Hold up 1 ring so that its eye hook is aligned in between the 2 eye hooks of the drop point. Thread the cotter pin attached to the shortest 1' length of string through all 3 eye hooks, so that the pin holds

the ring up, and if you pull on the string, the ring will fall.

STEP THREE: In the same way, hang the second hoop on the middle drop point with the 3½' length of string. Make sure the second string is above and out of the way of the first hoop; to ensure this, run the string up and over the clamp on the first drop point.

STEP FOUR: In the same way, hang the monkey target on the third, rear drop point using the longest, 6' length of string. Again, keep this string above the first and second hoops by running it over the clamps.

STEP FIVE: This is an important step. When everything is set up, unwrap the end of the long string attached to the stomper (which you did in STEP ONE). Then, carefully pull the string tight, ensuring there's as little slack in the string as possible, and rewrap it around the eye hooks on the stomper. It's tricky because you don't want to pull the cotter pins out and have your targets drop while you're adjusting the tension. The goal is to

blowgun (if necessary, use a short stepping stool to get up high enough to see through the ring). *Make sure no one is standing near the stomper while you do this!* One accidental stomp and you'd get hit in the face. We don't want anyone losing an eye!

When it's safe, look right down the barrel, and you should be able to see if the blowgun is aimed right at you. If it needs to go up or down, add or remove scraps of wood from under the front of the barrel. If it needs to move left or right, slide the barrel left or right. Remember: Move out of the line of fire each time adjustments are made to the blowgun!

Also, make sure your targets are lined up. If the boards of your frame are slightly warped or not squarely attached, you may need to line up your targets by moving the eye hooks or shifting the clamps.

Then, every time you shoot the blowgun, recheck your aim. We find that we get maximum success when we aim for close to the bottom edge of the rings rather than aiming for the center. See where you get the most success with your setup.

STEP EIGHT: Fire! With everyone safely off to the side or behind the blowgun, and out of the way of the dart, give the stomper a good, solid stomp. If all goes well, the dart will fly through both of the falling rings and hit the pie tin with a satisfying "thwap." It may take a few tries and some adjustments to get it perfect. After every attempt, check the tension on the strings and the aim of the barrel. If you find that your target is hitting the ground before the dart gets there, you may have to move your gun closer (and adjust your string tension accordingly).

How do you reinflate the stomped soda bottle? Before you put the dart back in the barrel, simply blow hard and fast into the end of the barrel, and the bottle will pop back into shape!

Once you get everything adjusted, it works beautifully. We've been able to hit the target 8 or 9 times out of 10. See how finely you can calibrate your setup!

have the pins pull out immediately as soon as you step on the stomper—not any sooner and not a moment later. With the string firmly wrapped around the eye hooks on the stomper, go to each cotter pin in turn, and, if needed, adjust the knot attaching it to the string so that all the strings have as little slack as possible.

STEP SIX: Load the blowgun. Insert the dart into the barrel. First, push in just the surveyor's tape tail so that it doesn't get wrapped around the dart, then push in the dart itself. It should slide in easily. Then use the ¾" dowel to push the dart down to the bottom of the barrel. Just push it down gently until you feel it stop. If you push too hard, it may compact at the bottom of the barrel and not fire as smoothly.

STEP SEVEN: Aim. The easiest way we've found to aim this straight at the target is to stand right behind the second ring and look straight at the barrel of the

THE SCIENCE

The dart hits the falling monkey. So where did Aristotle go wrong? Yes, a heavy object like a stone falls faster than a light object like a feather. But they don't fall at different rates because of their relative masses. There's something else at work.

What fooled Aristotle was air resistance. What he was observing was not the difference between heavy objects and light objects, but the difference between objects with lots of air resistance (relative to their weight) and objects with very little air resistance (relative to their weight). Feathers and leaves have shapes that create lots of air resistance when they fall. Sticks and stones, not so much. When air resistance is eliminated, everything falls at the same rate. In fact, there's a classic demonstration of this in which a penny and a feather are dropped at the same time—inside sealed tubes from which all the air has been pumped out. With no air resistance to slow the feather down, it falls just as quickly to the bottom and at exactly the same speed as the penny.

In 1589, the Italian scientist Galileo (who also makes an appearance in "The Pendulum Wave Machine," page 150) discovered that objects near the Earth's surface fall with the same acceleration, no matter what their weight. The story goes that he dropped two balls with different masses off the Leaning Tower of Pisa at the same time . . . and they hit the ground at the same time.

So, let's go back to our simple version of this experiment, where the hunter aims straight ahead at the monkey:

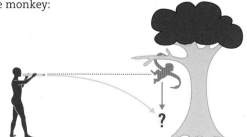

The monkey and the dart will fall toward the ground at the same rate. That means that, even though the dart is also moving horizontally toward the monkey, they will both fall the same distance toward the Earth in the same amount of time. Since they start at the same height, they will remain at the same height as they fall. That means the dart will hit the monkey (unless they both hit the ground first).

It's important to note that air resistance can still affect this experiment. The dart may fall more slowly than the monkey if the dart experiences more air resistance. If, for example, the dart were heavily feathered, the air would slow its fall. And if the dart were winged like a paper airplane, the monkey would certainly fall faster and escape. (For more on air resistance, see Paper Airplane Squadrons and Sharpshooters page 122.)

In this version of the experiment, the hunter isn't shooting straight ahead. We aim the blowgun up at an angle. The mathematics of the trajectory are more complex, but the result remains the same.

In addition, the string mechanism causes a tiny delay between the stomp and the dropping of the monkey, and the small tail on our dart causes a bit of air resistance. But if we're careful, neither of these has a big enough effect to derail the experiment. If we aim the barrel of the blowgun  directly at the monkey, the dart will fall at the same rate as the monkey, which will, most unfortunately, find itself darted despite its attempted escape.

If, however, you had an electromagnet release triggered by a laser and a more sophisticated launcher, you could shoot a ball bearing and hit another ball bearing as it drops. That's how precise this experiment can get.

 For more ideas, videos, and variations, visit www.eepybird.com/experiments/monkey.

With some water glasses, zip ties, a handful of screws, and a few tape measures, you can build a giant mechanical xylophone that plays any tune you like in your very own hallway! Two little mallets are pulled along the ground by a retracting tape measure, and as they slide by two rows of glasses filled with varying amounts of water, they tap out a melody on the glasses.

We start you out with a very simple tune, "Shave and a Haircut." This musical phrase has just a few notes. Then you'll play an entire song: "Twinkle, Twinkle, Little Star." We've picked these melodies because they're simple and involve a small range of notes. Once you see how it works, you'll be able to build your own version that plays any song you choose.

For the shorter version, you only need 7 glasses and a single tape measure; for the bigger version, you need 21 glasses and 4 tape measures. The glasses don't all need to be the same, so try to rustle up enough glasses for the full song if you can—dig around the basement and scrounge at yard sales. It's really impressive!

HOW DOES IT WORK?

Most everyone has made sounds or notes by blowing across the top of the mouth of a bottle. Then, as you drink more, the sound that the bottle makes changes. The same principle applies to the tone you get when tapping on a glass of water with a mallet: as you adjust the amount of water in the glass, you get different musical notes. Stephen once took a tray of glasses, put just the right amount of water in each one to make a complete musical scale, and then he used this array to play "Stars and Stripes Forever."

Playing that song required only a few glasses but a lot of practice. This experiment, however, requires more glasses and almost no practice (or musical ability). You will create not just one glass for each note in the scale, but one glass for each note in the *song*. Then, you place the glasses in the correct sequence, and it's automatic: the retracting tape-measure mallet taps the glasses in succession, and the melody rings out.

How do glasses of water make a musical instrument? Let's dive in.

THE EXPERIMENT: SHAVE AND A HAIRCUT—TWO BITS!

There are two parts to building this experiment: 1) the small mallets that are pulled by the tape measures, and 2) the water glasses that, when struck, vibrate and create the musical notes. No matter how long the song, or the number of glasses or mallets used, the principles remain the same.

MATERIALS

- one 12" length of 2 x 4 lumber
- four 2" drywall screws
- 2 zip ties, approximately 4" long
- duct tape
- one 25' metal tape measure
- one 4" x 4" piece of scrap cardboard
- 7 tall dinner glasses, all at least 5" to 5½" high (prefer sturdy glasses over thin, fragile glasses)
- water
- masking tape or sticky notes (for labeling the glasses)

TOOLS

- handsaw or chop saw
- screw gun or screwdriver
- scissors or utility knife
- piano, keyboard, or electronic tuner
- turkey baster (for tuning, optional)
- 1 work glove

HOW TO BUILD IT

Making the Mallets

The mallets consist of 2 metal screws attached to the ends of zip ties. These mallets are then attached to a wooden block at the end of an extended tape measure. When the tape measure is released, it rewinds and pulls the mallets along the floor. As the mallets slide past the water glasses, the mallers hit the glasses and play your tune.

STEP ONE: Using a saw, cut the 2 x 4 into three 4" lengths.

STEP TWO: Place one 4" piece on its edge, and lay another 4" piece flat on top, making a T (see illustration). Screw them together from the top using 2 screws. Turn the T upside down; this is the base for your mallets.

STEP THREE: Create the mallets. Take 1 zip tie and tape 1 screw to one end, making sure that the head of the screw remains exposed. (You can attach the screw to either end of the zip tie.) The screw head is what will strike the glasses to make a musical tone. Repeat with the other zip tie and screw.

STEP FOUR: Using the ends opposite the screws, overlap the zip ties about half an inch and duct tape them together. This makes a double-headed mallet that's about 8" wide. Then tape the double mallet horizontally onto the back and near the top of the upside-down, T-shaped wooden mallet base.

STEP FIVE: Using scissors or a utility knife, cut a 4" x 4" square of cardboard and tape it to the bottom of the mallet base. The cardboard will reduce the friction between the floor and the tape measure as well as protect the floor from scratches as the mallet base slides over it.

STEP SIX: Use duct tape to attach the tape measure (standing upright) to the remaining 4" piece of 2 x 4 (which lays flat underneath the tape measure), making sure not to obstruct the workings of the tape measure itself.

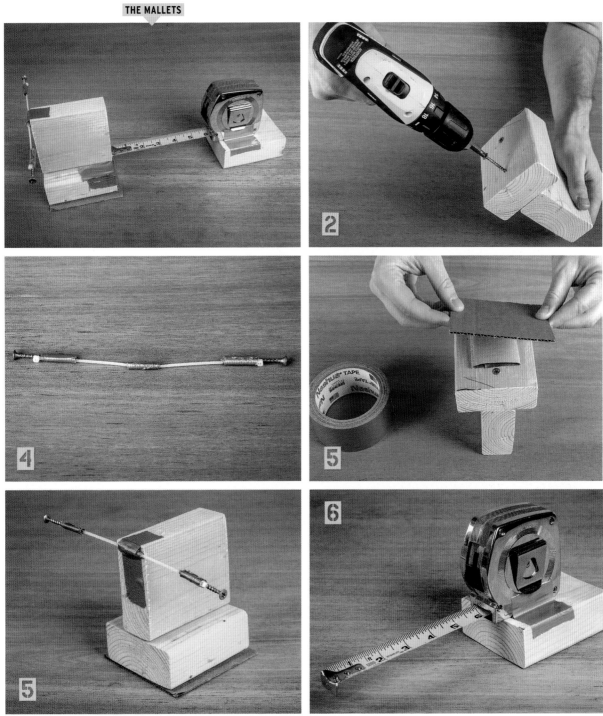

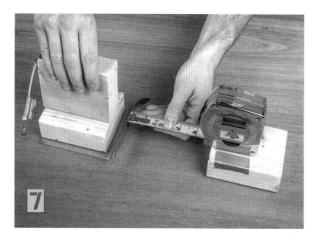

7

STEP SEVEN: Attach the end of the tape measure to the
mallet base. With the mallet base positioned so that the
mallets are facing away from the tape measure, extend
the tape measure and use duct tape to secure the tape
measure's lip to the back of the mallet base.

Later, when you're ready for the concert, you can secure
the tape measure by taping its 2 x 4 base to the floor
(making sure beforehand that the duct tape won't
mark the floor). This will keep the entire mallet-and-
tape-measure mechanism from shifting, both during
setup and while it operates; otherwise, if it shifts out of
alignment, it won't play correctly.

Tuning the Glasses

The Mechanical Water Xylophone can play any song
you want. However, each song requires tuning and
arranging the glasses properly. This mechanism is
sort of like a player piano, which only plays the tune
you tell it. This is the setup for "Shave and a Haircut,"
which is a surprisingly satisfying 7-note tune.

At this point, you need access to a piano, real or
electric, or some kind of electronic tuner that can guide
you as you tune each note. We like to use online tuning
sites or smartphone tuning apps. Visit EepyBird.com for
information on where to find these resources.

Don't worry, you don't need to be able to read music
or even be a musician to do this. You're just filling
glasses with water and then using your ears to adjust
the water levels to hit the right notes. In fact, for the
basic version, you only need 4 different notes, which
are shown on a piano keyboard in the illustration on
the next page.

To play "Shave and a Haircut," you'll need 7 glasses
tuned to these 4 notes:

→ C (3 glasses)   → E (1 glass)

→ D (1 glass)   → F (2 glasses)

Tune your first glass to C, or "middle C" on your
keyboard or piano. Fill the first glass with water until
it's approximately three-quarters full and tap the side
of it several times with a spoon, pencil, or a wood
mallet to sound the tone. While it's ringing, play the
same C note on your tuner or keyboard. If the note
your glass makes is too high, add some water, and try
it again. If it's too low, remove some water. Repeat this
process—comparing the notes, changing the water
level, and sounding again—until the two notes are the
same.

To speed up tuning, we recommend using a turkey
baster to add and remove water. You can use a spoon,
but a turkey baster is faster. As you fine-tune a note,
you will be adding or removing progressively less
water each time.

Essentially, the more water there is in the glass, the lower the note; the less water there is, the higher the note. However, the size and shape of the glass will also affect the note; if you have glasses of varying shapes, they may require different water levels to sound the same note. Even two seemingly identical glasses may differ. One huge benefit of the turkey baster is that it allows you to add or remove water very slowly while steadily tapping the glass and sounding the note at the same time. As you do, you'll hear the note slowly get lower or higher. Then, once you have one glass in tune, tune all the other same-note glasses, matching them to each other as well as to the tuner.

Marking Your Notes

As you tune each glass, label the note on a piece of tape or a sticky note and place it on the glass, aligning the top of the tape or sticky note with the exact height of the water level.

First, you need to keep track of your notes and distinguish your glasses. But also, this way you can empty your glasses and put them away without losing your tunings. Even after a day, enough water can evaporate to throw off your tuning. With your glasses labeled at the right spot for each note, you'll be able to return the water to the exact right level without having to use a tuner.

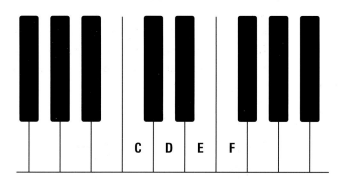

Notes for "Shave and a Haircut"

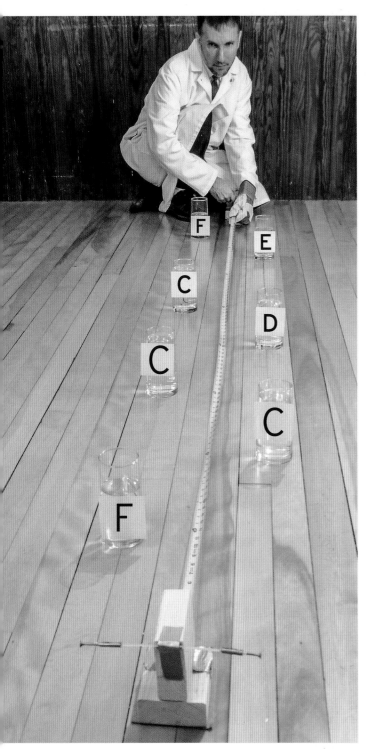

Lay Out Your Xylophone

Once the glasses are tuned, arrange them on a table in this order:

F C C D C E F

If you tap them from left to right, you should hear the classic melody, and you can sing along with the words: Shave and a hair cut—two bits!

If it sounds right, and no more tuning is necessary, you're ready to set up the mechanism. First, choose a long hallway or large space with a hard floor. Next, pull out the tape measure to 12 feet and lock it open with the tape extended. At this point, it can be helpful to tape the 2x4 block holding the tape-measure body to the floor, so it won't move.

Place your water-filled glasses on either side of the tape measure as shown in the photo to the left. How you space the glasses is how you set the rhythm, or the timing of the notes. Except for a single pause before the last two notes, you'll space the glasses evenly. The glasses should be far enough away from the tape measure such that the screw heads on the mallets will just strike the glasses as they slide by. You can use the tape measure itself to help space the glasses: place them 1' apart from each other, and after the first 5 notes, add a 2' double space for the pause before the final 2 notes.

When all of your glasses are in place, unlock the tape measure, let it rewind, and listen as the mallets play lovely music! If the music plays too quickly (if the tape measure retracts too fast), spread the glasses farther apart. Or you can slow down the speed of the retracting tape with your hand; put on a work glove, and apply some pressure with a finger as the tape retracts. The work glove protects your hand from the relatively sharp edge of the moving tape measure, which could cut your skin.

If the tape measure isn't very springy and pulls the mallets very slowly, you may need to extend it and pull it in a couple times to revitalize it. If it still isn't very springy, it may be time for a new tape measure.

With more glasses and more tape measures, you can play an entire song like "Twinkle, Twinkle, Little Star." The setup is more complicated, but the mechanism is the same (and uses all the same materials): for this song, you need 4 tape-measure mallets and 21 glasses.

"Twinkle, Twinkle, Little Star" has a total of 42 notes, yet like most songs, it repeats several sequences of notes, called musical "phrases." This repetition means that we only need 21 glasses in order to play the whole song, rather than needing 42 glasses, one for each individual note. In this set up, the mallets will hit each glass twice to construct the whole tune.

The song starts with this sequence of 14 notes:

> Twin-kle, twin-kle, lit-tle star. (pause)
> How I won-der what you are.
> C C G G A A G (pause) F F E E D D C

The next part consists of a different 7-note phrase, which is then immediately repeated:

> Way a-bove the world so high
> G G F F E E D
> Like a dia-mond in the sky
> G G F F E E D

Then the song finishes by repeating the same 14-note phrase that it starts with:

> Twin-kle, twin-kle, lit-tle star. (pause)
> How I won-der what you are.
> C C G G A A G (pause) F F E E D D C

This, then, is why you need 4 tape-measure mallets: 2 mallets will play the first phrase twice—once at the beginning of the song and once at the end—and 2 mallets will play the 7-note phrase twice in the middle of the song.

Here are the details:

STEP ONE: Construct 4 tape-measure mallets as described in the Shave and a Haircut experiment (page 182). If you've already made 1, then you need 3 more.

STEP TWO: Tune the 21 glasses, as described under "Tuning the Glasses" (page 184). For this song, you'll need the following number of notes:

→ A (2 glasses)   → D (3 glasses)   → F (4 glasses)

→ C (3 glasses)   → E (4 glasses)   → G (5 glasses)

STEP THREE: Lay out all 21 glasses and 4 tape measures like this:

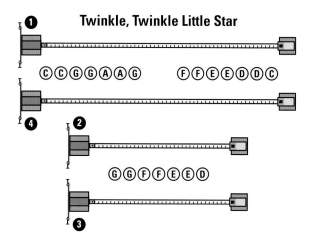

Twinkle, Twinkle Little Star

In order to play the same phrase on the same glasses twice, arrange the glasses for that phrase in a single line, and place a tape measure on either side of the glasses. To play the musical phrase twice, you retract first one tape measure (on one side) and then the second tape measure (on the other side).

STEP FOUR: Play it! Release tape measure #1 ("Twinkle, twinkle, little star. How I wonder what you are."). Then release tape measure #2 ("Up above the world so high."). Then tape measure #3, playing the same notes as #2 ("Like a diamond in the sky."). And finally, release tape measure #4, playing the same notes as #1 ("Twinkle, twinkle, little star. How I wonder what you are.").

Take a bow!

MEASURE FOR MEASURE: PLAYING MORE SONGS

With enough glasses and enough tape-measure mallets (and enough space), there is no song you can't play. As songs get more complex, the Mechanical Water Xylophone becomes a logistical puzzle, but it's musically very straightforward: you get one note per glass, and each line of glasses (or musical phrase) is capable of being played twice.

To get really low notes or much higher notes, you may need to use different-sized glasses, and you can also see what range of notes you can get out of glass bottles.

If you want to get really fancy, play a song with a little bit of harmony. Each tape measure has two mallets attached to it, so simply position two harmoniously tuned glasses on either side of the tape at exactly same spot, and the mallets will strike both glasses simultaneously. Then, adjust the spacing of glasses to create syncopated rhythms. Altogether, the Mechanical Water Xylophone allows you to play surprisingly complex music.

For added flourish, make a fifth tape-measure mallet, get 7 more glasses, and end the song with the "Shave and a Haircut" refrain. You're sure to get applause!

Sound is Vibration

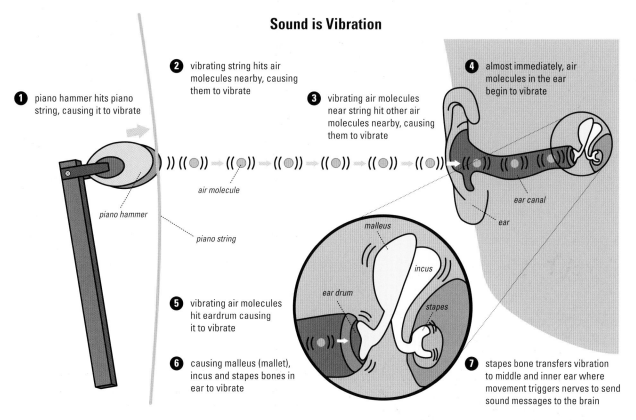

① piano hammer hits piano string, causing it to vibrate

② vibrating string hits air molecules nearby, causing them to vibrate

③ vibrating air molecules near string hit other air molecules nearby, causing them to vibrate

④ almost immediately, air molecules in the ear begin to vibrate

air molecule

piano hammer

piano string

ear canal

ear

malleus

incus

ear drum

stapes

⑤ vibrating air molecules hit eardrum causing it to vibrate

⑥ causing malleus (mallet), incus and stapes bones in ear to vibrate

⑦ stapes bone transfers vibration to middle and inner ear where movement triggers nerves to send sound messages to the brain

Sound is made of waves. Invisible waves. These invisible waves can travel through air, water, metal, plastic, or any other substance. Even though we can't see them, we can hear them—and sometimes when they're low and powerful enough, we can feel them.

To understand sound, you need to understand waves, and there are plenty of examples of visible waves that help us see what those invisible sound waves are doing. For instance, take waves in water.

Waves are one of the ways that energy moves from one place to another. A dramatic example is when an earthquake in the middle of the ocean creates a giant wave, or *tsunami*. With a tsunami, the energy of the earthquake moves through the seawater, and this wave of energy can travel for thousands of miles until something stops it, like land. For example, the 2004 Indian Ocean earthquake is one of the largest earthquakes ever recorded. It occurred off the coast of Sumatra, in Indonesia, and it created a tsunami that caused coastal flooding and deaths around the world, including as far away as Somalia, over four thousand miles from where the earthquake occurred.

In principle, at least, this is no different than if you poke your finger into a sink full of water. Your finger disturbs the water and sends little pulses of energy radiating out.

What happens, in both cases, is that the disturbed water pushes into the water around it, and that water pushes the water around it, which pushes the water around it, and so on. In water, we can see this moving energy embodied in the wave, which moves *through* water, lifting the water up and leaving it behind as the energy itself keeps moving.

SOUND WAVES

Sound is also a wave phenomenon; in this case, it's a result of vibrating air. When you press a key on a piano, that key causes a felt-covered mallet to hit

Wavelength Determines Pitch

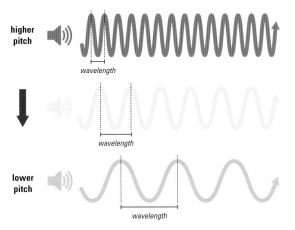

The shorter the wavelength, the higher the pitch.
The longer the wavelength, the lower the pitch.

a tightly stretched metal wire. As the wire vibrates, it vibrates the air around it, which creates sound, a particular note.

When the piano wire vibrates back and forth, it bangs into the air molecules nearby; those molecules bang into the molecules around them, and so on, creating a wave in the air just like a wave in the water. When the wave finally reaches the air molecules next to your eardrum, those vibrating molecules make your eardrum vibrate, and that's what allows you to hear the note.

Wave Height Determines Volume

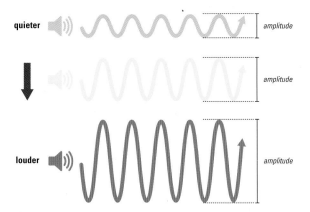

The greater the amplitude (wave height),
the louder the sound.

Each wire in a piano is designed and tuned to produce a different vibration. Wires that vibrate faster produce notes we hear as higher pitches. Wires that vibrate more slowly create pitches that we hear as lower notes.

Waves can vary in two main ways: how fast they are, which is known as the "frequency" of the wave, and how tall they are, which is known as the "amplitude" of the wave. The *faster* waves are, the higher their frequency, and the higher the pitch. The slower waves are, the lower the pitch.

The wave's amplitude (or height) affects how loud or soft it sounds. The *taller* the wave is, the louder the sound; the shallower the wave, the softer it becomes. The various combinations of these properties create the variety of sounds we hear.

To tune our water glasses for the Mechanical Water Xylophone, we take one more thing into account. The size or mass of an object affects the speed at which it vibrates. If you look at the strings inside a piano, you'll notice that thick strings make the lowest notes and thinner strings make the highest notes. The tension of the strings also affects their pitch, but if tension is relatively equal, thick strings will always vibrate more slowly and create lower notes.

In this experiment, we control the speed of the vibration, and the pitch of the note, by varying the amount of water in each glass. When you strike a water-filled glass, both the water and the glass vibrate together, and that vibration creates the tone. Glasses with more water have more mass, and so they vibrate more slowly and make the pitch lower. As you remove water, the mass decreases, the vibrations get faster, and the pitch goes up.

For more ideas, videos, and variations, visit www.eepybird.com/experiments/watermusic.

ACKNOWLEDGMENTS

This book would never have been possible without the help, encouragement, and advice of a lot of talented people. In particular, we'd like to thank Stephen's brother John Voltz, who jump-started this project, almost single-handedly propelling it from an idea to a book proposal to a sit-down meeting with exactly the right publisher, all in a matter of weeks.

In the laboratory, our team of EepyBird Inventors has been indispensible. Michael Miclon, Shane Miclon, Collin Miclon, Aaron DeWitt, Casey Turner, and Brian Miclon helped us research, build, and field test these experiments. They have inspired us, kept us laughing, worked hard, and pushed us to go ever further.

Thanks to our EepyBird photo team, working on both sides of the camera: Mike, Shane, Collin, and Kristen Phillips.

And special thanks to:

Emily Haynes, our editor at Chronicle who saw our book proposal, understood immediately what this book could become, and got behind it from the beginning.

Jeff Campbell, our fantastic copy editor, who provided invaluable help refining the final versions of the manuscript.

Project manager Kate Willsky, art director Neil Egan, designer Alissa Faden, and illustrator Hillary Caudle, for taking our manuscript, photos, and sketches and turning them into this beautiful book.

Stephanie Kip Rostan, our literary agent, who has guided us so well.

Thanks also to William Beaty for his help with DIY hovercraft history and the science of nucleation.

And, of course, to Julia Brotherton and Charles & Elizabeth Grobe for all their advice, support, and encouragement.

INDEX